£ 7.50
L-V
6/45

SIMONE de BEAUVOIR and JEAN-PAUL SARTRE

SIMONE de BEAUVOIR and JEAN-PAUL SARTRE

The Remaking of a Twentieth-Century Legend

Kate Fullbrook

and

Edward Fullbrook

BasicBooks
A Division of HarperCollinsPublishers

Library of Congress Cataloging-in-Publication Data
Fullbrook, Kate.
    Simone de Beauvoir and Jean-Paul Sartre : the remaking of a
twentieth-century legend / Kate Fullbrook and Edward Fullbrook.
        p. cm.
Originally published: Hemel Hempstead : Harvester Wheatsheaf, 1993.
Includes bibliographical references and index.
    ISBN 0–465–07827–3
    1. Beauvoir, Simone de, 1908– —Biography. 2. Sartre, Jean Paul,
1905– —Biography. 3. Authors, French—20th century—France—
Biography. I. Fullbrook, Edward. II. Title.
PQ2603.E362Z684   1994
840.9'00914—dc20
[B]   93–35968
                                                                        CIP

94 95 96 97 RRD 9 8 7 6 5 4 3 2 1

To: The Other

Vous faites un film d'entretiens, comme Sartre, vous publiez vos "Ecrits", comme Sartre . . . Vous n'en avez pas un peu assez d'être tout le temps comparée à lui?

Simone de Beauvoir: Etant donné la misogynie bien connue des gens, et en particulier des Français, il est vrai qu'on m'a toujours considérée avant tout comme la compagne de Sartre. Alors qu'il n'est jamais venu à l'idée de personne de considérer Sartre comme le compagnon de Simone de Beauvoir.

"Beauvoir elle-même"

CONTENTS

ACKNOWLEDGMENTS

We wish to express our thanks to Jackie Jones, Charles Butler, Geoffrey Channon, Tina Coulsting, Jean Grimshaw, Adolphe Haberer, Sue Habeshaw, Christine Huggins, Robin Jarvis, Chris Parker, Derrick Price, Renee Slater, Helen Taylor, and Janet Ward for help of various essential kinds. We are particularly grateful to Susan Manning for her help in commenting on the manuscript of this project when a less patient reader would have been defeated by its disorder. No literary project of any kind can emerge without the assistance of librarians, and, in this instance, we wish to thank the librarians of the St Matthias Library of the University of the West of England, Bristol (with special thanks to Glennis Kilbey) for their kind and helpful work on our behalf.

We also wish to thank the following individuals, estates, and publishers for permission to reproduce material in copyright from the works indicated. Every effort has been made to trace copyright holders, but if any have been inadvertently overlooked the authors and publishers will be pleased to make the necessary arrangements at the first opportunity. Jean-Paul Sartre, *The Words*. Copyright © 1964 by Librairie Gallimard. English translation copyright © 1964 by George Braziller, Inc.. The estate of Simone de Beauvoir for Simone de Beauvoir, *Letters to Sartre*, trans. Quintin Hoare (Radius, 1991). From *Letters to Sartre* by Simone de Beauvoir, Copyright © 1990 by Éditions Gallimard. English Translation Copyright © 1991 by Quintin Hoare. By permission of Little, Brown and Company. Reprinted by permission of Deirdre Bair, Copyright 1990 from *Simone de Beauvoir: A Biography*, Jonathan Cape, London. Simone de Beauvoir, *She Came to Stay*, trans. Yvonne Moyse and Roger Senhouse

INTRODUCTION

We always intended, since the inception of this project, to devote our introductory remarks to our experience of the process of collaboratively writing a study of one of the most famous collaborations of the century – the extraordinary personal and intellectual partnership between Simone de Beauvoir and Jean-Paul Sartre. We had developed, separately, long-term interests in the couple and in their work, and though the nature of our interests varied, we were intrigued to see what would happen when we combined our fascination, knowledge, and, at times, annoyance with well-known aspects of our subjects' life histories. At the beginning of our research, we felt we were treading well-known ground. After all, Beauvoir, in her four volumes of autobiography and in her studies of her mother's and Sartre's deaths, had provided enough material for any biographical undertaking. Sartre, like Beauvoir, was also the author of a classic autobiography, *The Words*, and the highly revelatory *War Diaries*. It seemed clear to us that the couple had themselves left more information than we could possibly use within the confines of a relatively brief study. In addition to their autobiographies, there were their many articles, travel writings and interviews on which to draw. It seemed to us, for a time, that Beauvoir and Sartre could scarcely have had time to live: they must have been too busy narrating their own experiences to so much as eat without pens in their hands or tape recorders at their elbows. They seemed to have specialized in telling all . . . repeatedly.

Further, we soon became aware of just how well the couple had been served by their biographers. It is the most obvious of truisms to note how serious research concentrates attention, and we soon realized just how much ground the dedicated biographers of Beauvoir and Sartre, whom we had read casually, had covered. We very much want to note our debts to the pair's earlier biographers and interviewers whose work has necessarily been important, in a variety of ways, for us. Beauvoir has been fortunate to attract such scholars as Claude Francis and Fernande Gontier, Deirdre Bair, Margaret Crosland, and Alice Schwarzer who have all done much to attract readers to one of the most interesting writers of the century. Fine work has also been done on Sartre's life. We have learned a great deal from Annie Cohen-Solal, Ronald Hayman, John Gerassi, and from the superb bibliographical studies of Michel Contat and Michel Rybalka. All these biographies are dependent, as is only right, on the material provided by Sartre and especially Beauvoir about the couple's life and times. We, in turn, have debts to pay to all our predecessors.

Yet as our study proceeded, we became more uneasy with our, and with previous readings, of Sartre and Beauvoir's relationship. Chronologies did not seem to quite fit; comments made by Beauvoir in various places did not correspond to each other; Sartre seemed, especially in interviews, to go silent at moments when speaking should have been easy. Most of all, we became alert to the significant moments when Beauvoir gave warning, in various ways, that while she intended, in the main, to tell the truth about herself, Sartre, and their associates, she felt no compunction (as was her right) about changing things, hiding things, letting discretion prevail wherever she felt the need. Further, our repeated readings of Beauvoir's autobiographies – a main source for all biographers of the pair – revealed just how tightly *structured* they were. No life is so neat. That such a life had been proposed as her own by a novelist who had the talent to sustain a massive international readership since the 1940s began to indicate to us the need for caution.

During our period of research for this study there were striking developments in the appearance of documents dealing with the couple's lives. Beauvoir's *Letters to Sartre* and *Journal de guerre* confirmed the bisexuality she had been at such pains to hide from the public throughout her life, even in the period of the last decade before her death when many of her feminist supporters would have welcomed such revelations. Further, these new developments confirmed the fact that her sexual activity with both men and women was symmetrical with Sartre's. The classic view of Beauvoir, encouraged by her own writing and by Sartre's acquiescence, was one of Sartre as a womanizer and Beauvoir as the patient, loyal female victim (in as much as anyone as forceful as Beauvoir was likely, in maturity, to be anyone's victim). This view was clearly, and provenly, no longer tenable. For our part, we were less interested in the pair's sexual arrangements, though biographers cannot help but be interested in this kind of gossip (even when it is fifty years old), than in Beauvoir's motivations for maintaining her constructed and very different identity for the couple with such persistence over so many years. Previous biographers and interviewers had clearly been fed a line – a lot of lines – for reasons that were difficult to understand completely.

What concerned us most were issues related to intellectual indebtedness and the development of ideas within the confines of the lifelong, and clearly not transparent, relationship between the couple. By slowly working through the chronologies of their writings and using the new material that has become available both to note newly revealed facts and to reread the old information in the light of the new, we have come to some startling conclusions.

It is now utterly clear to us that Beauvoir was always the driving intellectual power in the joint development of the couple's most influential ideas. The story of their partnership has been told backwards. Our detailed work on the genesis of *She Came to Stay* shows, incontrovertibly we believe, that the major ideas behind *Being and Nothingness* were fully worked out by Beauvoir and adopted by Sartre before he even began his famous study. The relationship between these two key works forms the heart of this study, and it is probably fair to say that we are as surprised as any reader might find themselves by what we have found. We expected, in the beginning, to need to work, like Michèle Le Doeuff, to adjust notions of the relative sophistication of Sartre and Beauvoir's thought. We did not anticipate that our findings would shift the balance so radically and decisively in Beauvoir's favour.

In the process of rereading the history of this most famous of twentieth-century literary partnerships we have been startled, baffled, angered, and delighted by the often tortuous complications of the very complex association of our subjects. What we are clear about, however, is that their legend is just that, a legend, constructed, for reasons that must forever remain partially unknown, by the couple for their own purposes. The truth lies elsewhere. We hope that this volume, the first of two, will contribute something to getting a little closer to it.

1

THE PRECOCIOUS PLAGIARIST

In his old age, Sartre became careless about the legend he and Beauvoir had so carefully constructed around themselves. On March 26, 1971, in an interview with John Gerassi, he mentioned that he had not been Beauvoir's first lover, and named as her first sexual partner a man still living who, when interviewed, confirmed the revelation.[1] It was not the fact itself that was startling, but rather the implications it might hold for the credibility of Sartre and Beauvoir's many-volumed and often-told version of their connected lives. Her sexual innocence was integral to the most hallowed of all Sartre-and-Beauvoir stories, that of the initial meeting and romance, which culminated in their oath-taking in the shadow of the Louvre. If *this* story was based on a fabrication, then so too might be any of their others. And what did this suggest about the *real* nature of the relationship that existed between them? What was it that was being hidden? Surely it was not simply the fact of Beauvoir's lack of virginity when she first had sex with Sartre. Sartre's perhaps unintentional revelation pointed to something dubious about the legend he and Beauvoir had invented about themselves, something that had to do with the terms on which they had founded that relationship.

In another of the Gerassi interviews, which for reasons beyond the interviewer's control remain unpublished, Sartre offered the following information about his grandmother and the part she played in his intellectual development when he was 15: ". . . she at seventy-seven, alert as ever, reread and hence impressed me to read Dostoyevski, Tolstoy, Stendhal, so we could discuss them. I spent hours with her. We even discussed the Russian revolution . . .".[2] What is startling here is not the unexpected range of interests of Sartre's grandmother, but the radical

4

way this disclosure undermines the reliability of *The Words*, his celebrated account of his early childhood. Jean-Paul Sartre was 15 months old when his father died and from then, until the age of 12, he and his mother lived in the household of her parents. *The Words*, the most carefully crafted of all of Sartre's books, is the story of those eleven years and of the parts played by the three adults in his formation as a future writer. Because he was, in the main, kept out of school until the age of 10, and did not play with other children, his mother, grandmother and grandfather together had virtually total control over him, a fact that did not sit well with the adult Sartre's well-known preference for thinking of himself as self-created. In his autobiography, he deals with this by drawing attention away from the two women and toward his grandfather, Charles Schweitzer, whom he identifies with learning and portrays as a larger-than-life patriarch who dominates his wife and daughter to the point of reducing them to insignificant beings. The wise grandson, however, even at the age of 6, sees through the pomposity of the old man and thus ultimately escapes his influence. It is a measure of Sartre's skill as a narrative writer that he largely succeeds in making the reader believe his story.

The Words was published in 1963, when Sartre was in his late 50s and when most of his significant works had already been published. In retrospect, it is clear that one of Sartre's chief objectives in his autobiography is to draw attention away from the decisive effect that women had on his intellectual formation and to secure for himself a masculine intellectual lineage.

In *The Words*, it is Sartre's grandfather who is given roles of educational initiator, antagonist and prime force in the shaping of the young Sartre's mind and principles. Sartre handles the narrative of his young life in schematic, indeed folkloric terms. As Sartre tells his story, his grandfather, Charles, a provincial school teacher who had written two books of arcane historical scholarship and textbooks for teaching foreign languages, had been about to retire when his daughter Anne Marie returned home widowed, penniless, and with a child. A few years later, having reached the mandatory retirement age, Charles moved his family of four to Paris where he successfully set up a school for teaching French to foreign students. At 70, says Sartre, Charles was still agile, haughty, tall, handsome, had a booming voice and a long flowing beard stained with tobacco, and "so resembled God the Father that he was often taken for Him."[3] Repeatedly in *The Words*, Sartre offers strong visual images of Charles, whereas the two women are never allowed to materialize in the reader's eye. About his grandmother, Louise, of whom he must have seen a great deal more than his working grandfather, Sartre says remarkably little, nearly all of it meant to dampen interest in her. Sartre explains that Louise did not enjoy sex with Charles, and that she raised their three children as Catholics against Charles's opposition. She was, says Sartre, "shrewd but cold," "thought straight but inaccurately," "read lots of spicy novels," and was reclusive to the point of remaining behind her bedroom door for days at a time.[4]

Sartre's mother, with whom he lived alone during most of the ten middle-aged years he worked intermittently on his autobiography, presented him with a much more serious narrative challenge than his grandparents. He knew both that the closeness of his childhood relationship with his mother could not be denied and that the pattern of his adult life, particularly if the story of his stepfather became known, invited an Oedipal interpretation. To have, in this way, been denied his claims to self-generation would have been especially humiliating after his decades of writing against Freudian reductivism. Sartre's solution was to risk all on a daring stratagem. He would offer a full-scale, full-color rendering of the mutual devotion between mother and child, including the former's encouragement of her son's early and inordinate proclivity for writing, but he would psychologically neutralize the rendering by presenting his mother as a sister-figure.

Sartre could make his presentation of his mother as his sister plausible because there was an element of truth in it. Being a woman in early twentieth-century France meant that Anne Marie's legal rights were scarcely greater than her toddler son's, and, also like Jean-Paul, she was materially dependent on her parents, given no money, and allowed out only with permission. Mother and son even slept in the same room, called "the children's bedroom." Throughout *The Words* Sartre reinforces the idea that he perceived his mother as a sibling by nearly always referring to her as Anne Marie or as his "room-mate," but his descriptions of their everyday existence reveal him as an archetypal mother's boy. He recalled his beloved's sugared injunctions to him: "My little darling will be very nice, very reasonable. He'll sit still so I can put drops into his nose." Anne Marie's prophecies of obedience came true. Sartre presents himself as falling in with her wishes: as a child he was well behaved, never cried, seldom laughed, and knew "nothing more amusing than to play at being good." Supremely contented with his lot, Sartre later attributed this happiness largely to the absence of a father. "My mother was mine; no one challenged my peaceful possession of her. I knew nothing of violence and hatred; I was spared the hard apprenticeship of jealousy."[5] For the next ten years, says Sartre, Anne Marie devoted herself to almost nothing but her contented son and room-mate.

But it was not only Anne Marie who is shown as doting on Jean-Paul in *The Words*; the formidable Charles Schweitzer is given the same role. "He would call me," recalls Sartre, "his 'tiny little one' in a voice quavering with tenderness. His cold eyes would dim with tears. Everybody would exclaim; 'That scamp has driven him crazy!' He worshipped me, that was manifest." The child responded by playing to the old man's weaknesses, joining him in elaborate charades of mutual affection and admiration. Charles insisted that he, himself, had been a child prodigy and claimed the same status for his grandson. Just as Sartre obliged his mother with an imitation of obedience, so, too, he invented himself as the prodigy Charles Schweitzer demanded. Jean-Paul parodied the adults' speech with great success; his preciosity was applauded and constantly discussed. His mother and grandfather, for their different reasons, openly idolized the "tiny little one" until he came to see himself as the embodiment of the Good and the True in an almost perfect world, characterized by Progress, which culminated in him. The flaw in Jean-Paul Sartre's paradise was his grandmother:

> . . . it pained me to note that she didn't admire me sufficiently. In point of fact, Louise had seen through me. She openly found fault with me for the hamming with which she did not reproach her husband: I was a buffoon, a clown, a humbug; she ordered me to stop "smirking and smiling."[6]

Sartre's account of the upheaval that invariably followed Louise's puncturing of the family's charades of perfection and mutual adoration shows that the inner workings of the Schweitzer household, far from being simply patriarchal, as Sartre overtly claimed, were complex and convoluted. Little Sartre would counter his grandmother's orders. Louise would demand an apology. Jean-Paul would refuse. Charles would side with him; Louise, outraged, would storm away to her room. Patriarchal power would seem to triumph. But Anne Marie, who, Sartre says, feared her mother, would accuse Charles of causing the quarrel. Schweitzer, in turn, would follow his wife's example and lock himself away. Anne Marie would "beg" her son to ask his grandmother's forgiveness. In the end, it was Sartre's mother's authority that weakly prevailed, and, playing his role of the perfect child, Jean-Paul would go to Louise and "apologize casually."[7] Despite their seeming capitulation to Charles's control, Sartre learned early that the women of his family were both sharp-witted and powerful. It was a lesson that was to affect his relationships with women throughout his life.

At an early age, Sartre observed that books were among the most favoured of props in the Schweitzer soap opera. There were two kinds: grandmother's and grandfather's. Louise read two thick popular novels a week and exchanged them at the library every Friday, a process that mystified her 3-year-old grandson, as did the conversations between mother and daughter about these dense objects at which the adults stared for hours on end. On Sundays, Charles, who would not read a novel written after the death of Victor Hugo, would enter his wife's room, take the novel she was reading from her hands, read a few lines, declare "I don't get it," throw down the book and stalk off to his book-lined study. There too, the still illiterate Sartre was "a daily witness of ceremonies whose meaning escaped" him. Without understanding their function, he learned to revere his grandfather's mysterious "boxes," over a thousand of them, that "split open like oysters."[8] Sartre was particularly awed by the textbooks Charles himself had written, which were pointed out to him with pride. It was made clear to the infant Sartre, before he could read, that books were sources of pleasure, pride, disputation, and power.

When his mother started reading rather than telling him stories, Sartre began to understand what books were about, and was duly impressed and soon jealous of his mother's role as reader. He contrived to be caught in the act of pretending to read a book entitled *Tribulations of a Chinese in China*. The family decided it was time to teach the boy the alphabet. As a child marooned and lonely in a family of doting and highly literate adults, it was not long before Sartre, aged 4, was reading too.

Losing the sight of his right eye at about the same time as he learned to read seems to have made surprisingly little impression on Sartre. More significant was his family's move, in 1911, from the small town of Meudon to central Paris to a high-ceilinged, sixth-floor flat located close to the Panthéon, the Sorbonne, and the Luxembourg Gardens where Sartre's mother now took him for walks. The one-eyed, pampered prodigy was only 5 and still had his baby curls, but, like a peasant, he had already found the earth in which he was to be rooted. Within a mile and a half of the Schweitzer flat at 1 rue le Goff, he would receive nearly all his education, live most of his life, die, and finally share a grave with a woman born and raised on the other side of his neighborhood park.

At first, Sartre was not sent to school. Sequestered in his grandfather's flat high above the city with only adults for company, Charles's library became Sartre's unlikely playground. Its French and German classics were beyond Jean-Paul's pretensions, but not its *Larousse Encyclopedia*, from which he would pick out a volume at random, position it, with effort, on Charles's desk, and explore the "flora and fauna, cities, great men, and battles" of *A–Bello*, or *Mello–Po*, or *Ci–D*. In this platonic and play-acting fashion, in which everything was labeled and classified, and in which he pretended to be his grandfather, Sartre first surveyed the universe.[9]

The adults applauded his bibliophilia, discussed Jean-Paul's "thirst for knowledge" in his presence, and asked him daily about what he had read, what he had understood. Charles, as high priest of learning, taught him the names of the most illustrious authors, "the Saints and Prophets" of the intellectual heavens. In the library – the "temple" – these divinities

waited to become Jean-Paul's playmates. He quickly learned that playing
with those judged the most sacred of cultural icons earned the most
applause, so he gravitated to Corneille – "a big, rugged, ruddy fellow
who smelled of glue and had a leather back" – even though Corneille's
Alexandrines remained unfathomable to him. The future biographer
of Flaubert memorized the last pages of *Madame Bovary*, but without
understanding Charles Bovary's behavior. In all this Sartre recognized
the depths of his own deception: his grandmother was right – he was a
fraud. "Even in solitude," confessed Sartre, "I was putting on an act."[10]
As he "read" Charles's holy works, he imagined the adults' approving
gaze upon him as he pretended absorption in literary works far beyond
his comprehension. It was a solemn game for a little boy, yet, in the long
term, this play-acting cultivated him as he began to find his way in the
"Saints'" jungle of words.

If Charles is treated by Sartre as the encouraging stern guardian of
high culture, he credits his mother with succeeding in the lighter task
of introducing him to the popular artifacts appropriate for a boy of his
age. Sartre insists that his mother increasingly became concerned about
the overly illustrious intellectual company her son was keeping, and,
secretly at first, introduced him to comics and children's books as an
antidote to premature solemnity. Overnight, without losing his appetite
for the classics or for the praise they brought him, Jean-Paul became
hooked on pulp fiction. Just as Louise lived for the Fridays on which she
exchanged her library books, Jean-Paul began to live for the Thursdays on
which new instalments of the comics in their garish covers appeared on
the newsstands. In due course Charles uncovered his grandson's literary
slumming and was enraged by it. In the ensuing family row, Anne Marie
and Louise defended Jean-Paul's right to his childish reading matter and
prevailed over the patriarch. Sartre remained a fan of popular fiction
throughout his life, a taste dignified by its association with victory over
patriarchal control, and made pleasurable through its association with
women.

This victory was followed by another, again engineered by Anne Marie,
who introduced Sartre to the delights of the cinema. He was enthralled –
especially by westerns – and went to the cinema regularly with his mother.
The literary and cinematic encounters with the adventurous fantastic were
of great importance to Sartre, and the delight they offered was fed by the
fact that Charles was passively dismayed to see his precious grandson
being raised by his mother on a steady, and all-too-welcome, diet of
comics, thrillers, and films.[11]

In *The Words*, Sartre pays particular attention to the moments of division
among the adults in the Schweitzer household as the means for shaping
his identity. Besides the choice of his reading matter and his attendance at

the cinema, a major bone of contention between Charles and the women of the house was the style of Jean-Paul's hair. It had not only been kept long: Anne Marie and Louise enjoyed curling it and using its grooming as a means of tender care for little "Poulou" (Sartre's nickname, which his mother continued to use for him until her death). Charles's reaction to this fuss was one of stereotypical masculine homophobia: " 'You're going to make a girl of him. I don't want my grandson to become a sissy!' " Unable to force the women to arrange a haircut for his 7-year-old grandson, Charles decided to oversee the operation himself and took Sartre off to the barber. Everyone, except the boy, was shocked by the result. Sartre's long curls had concealed his considerable ugliness. Charles, having left the house with a becurled wonder-child, "had brought back a toad."[12] At the sight of her shorn son, Anne Marie shrieked and ran into her room to cry. Even Charles was chastened.

Sartre's appearance played an important part in his sensitivity to his physical estimation of himself and in the shaping of his personality. As a little boy he had been told, and, like the adults, had believed, he was good-looking. He claims that it was another five years before he discovered the truth of his own ugliness. But from the time of his haircut onwards, "I had a general feeling of uneasiness. I would often catch friends of the family looking at me with a worried or puzzled expression. My audience was getting more and more difficult. I had to exert myself."[13] Exert himself he did, with the result that he developed uncommon powers to please others by the force of his personality. One of the preeminent characteristics of the adult Sartre was his studied cultivation of great charm as compensation for his equally extreme ugliness.

About a year after the haircut, Charles put another family belief at risk by enroling his grandson in school. Sartre had reached the age of 8 without ever setting foot in a classroom, and when, after a few hours, the principal of his first school reported that the "genius" was unable to keep up with the other pupils of his own age, Charles erupted. The next day he simply withdrew Sartre from school. The child was accused of not trying hard enough. Sartre claims he was unaffected by this absolute failure ("I was a child prodigy who was not a good speller, that was all") and was pleased to return to his solitude. Charles bought him a desk and hired a tutor to give him private lessons. This plan for a private education also failed; the tutor soon disappeared. The adults were puzzled about how to proceed.

In the following year Anne Marie attempted to initiate Sartre's formal education. Rather than exposing him to the rigors of an ordinary French classroom, she discovered what seemed a good option for her tender one. She found a school in which the mothers of the pupils not only accompanied their children to the premises, but remained with them in the classroom. Sartre remembered that the main duty of his teacher was "to distribute praise and good marks equally to our class of prodigies," but, apparently, the rewards distributed to Jean-Paul were too modest, because at the end of one semester he was withdrawn from this school too.[14] Subsequently, a succession of female teachers gave him private lessons at home.

At about the same time as the adult Schweitzers were staging their histrionics over young Sartre's appearance and education, he was secretly agonizing over a problem of his own. Living six floors above the streets of Paris, having no contact with his peers, and with no teachers allowed to criticize him, Sartre had only three adults and their compliant friends against whom to develop his sense of self. He had been successfully conscripted into the family drama and played his part of obedient, grandfather-worshipping child prodigy well. He threw himself into the assigned role unreservedly, and often with verve. But the future playwright's exuberance in acting out his prescribed part also left him feeling hollow:

> I was an imposter. How can one put on an act without knowing that one is acting? The clear, sunny semblances that constituted my role were exposed by a lack of being which I could neither quite understand nor cease to feel.[15]

This "lack of being," he says, came from his inability to see himself except through the eyes of Anne Marie, Louise and Charles, so that even when none of the triumvirate were present he would imagine himself still in character and under their gaze. With his identity based on how he imperfectly perceived himself as seen by three adults, his unease was much heightened when, at the age of 7 or 8, he realized that his role was a "false" one in "that though I had lines to speak and was often on stage, I had no scene 'of my own,' in short, that I was giving the grown-ups their cues."[16] To his dismay, Sartre discovered that he was not only playing the role of a child, but also really was one.

Jean-Paul's realization that he "did not really count" made him, like most children, want to grow up. But here, too, turning to the adults for help, he initially found a deficiency of being. He was told he would not follow his father into the navy, but, other than that, says Sartre, "nobody, beginning with me, knew why the hell I had been born." His ghostliness even seemed apparent to other children. When Anne Marie began taking him to the Luxembourg Gardens to play with other children, they ignored him. In a short time the child who had imagined himself a star came to feel superfluous, left out of both the present and the future. Sartre felt radically devalued. This awakening seems to have been genuinely traumatic for him. As an adult, even more than most people, Sartre was obsessed with the need to feel wanted by the people around him. Meanwhile, still proud, the 7-year-old looked for anything that might save him from nothingness, and found half of what he needed when he read *Michael Strogoff*. Its eponymous hero receives a command from on high which turns into a mission designed to give him a predestined life ending in glory. The idea that "certain individuals were chosen," with "their path laid out for them by the highest necessities" was new to Sartre, and, given his predicament, irresistible.[17] Almost overnight, he became a boy in search of a mission.

Literature, says Sartre, provided the opening. He was 7 and on holiday when he received a letter in verse from Charles. Anne Marie, amused, taught little Poulou the rudiments of prosody, and she and Louise helped him compose a versified reply. Charles responded in kind, and for a time this exchange was repeated thrice-weekly. Back in Paris, the budding poet attended catechism classes and wrote an essay on Jesus that won a silver-paper medal. Again, Anne Marie intervened with imagination and used this small success to turn her son's interest further toward his own writing. Thirty years later and a soldier at the front, Sartre recalled the importance of this event in his childhood:

> I am still filled with admiration and delight when I think of that essay and that medal, but there's nothing religious about this. The fact is my mother had copied out my composition in her beautiful hand, and I imagine the impression that seeing my prose transcribed in this way made on me was more or less comparable to the sense of wonder I felt at seeing myself in print for the first time.[18]

After his essay's success he decided prose was more to his liking than poetry, and, equipped with a notebook and a bottle of purple ink, he began translating his adventure comics into written stories. Being a writer was a "game" that the lonely child could play by himself.

In *The Words*, Sartre stresses the importance of his plagiarism, and describes it as multifaceted. Not only were all his characters, plots, and incidents lifted from his pulp reading, but when his shipwrecked hero cried "Shark!", Sartre would open the *S* volume of the *Larousse* and copy into his narrative the first paragraph of the entry for "Shark." He did the same for "Brazil" when his hero swam ashore. Despite all these clumsy borrowings, Anne Marie was

> lavish with encouragement. She would bring visitors into the dining-room so that they could surprise the young creator at his school-desk. I pretended to be too absorbed to be aware of my admirers' presence. They would withdraw on tiptoe, whispering that I was too cute for words, that it

was too-too charming. . . . Anne Marie copied out my second novel, *The Banana-seller*, on glossy paper. It was shown about.[19]

Charles refused to read it; the old family quarrel regarding elite and popular education was rekindled when Charles discovered that his grandson's writing was based on his "unwholesome reading-matter." Angrily, he refused to have anything to do with it, but this failed to deter Jean-Paul, who increasingly saw his writing as a means of "escaping from the grownups."[20]

In the family quarrel over the making of Jean-Paul the writer, Sartre gives his mother the role of angel when she decides that her son's penchant for writing is a sign of a natural vocation. Helping him into his nightshirt, she declared: "My little man will be a writer!" When she informs her father of the destiny she has selected for her son (an act of courage, because by this time Charles had decided his grandson would follow in his pedagogic footsteps), a violent outburst is expected. Perhaps sensing the delicacy of the situation, Charles postponed his response. If he had opposed the writerly idea directly, says Sartre, "I might have stuck to my guns obstinately."[21]

But Sartre portrays Charles's approach as wily. One evening he asked the women to leave the room. He wanted to talk to Sartre "man to man." Charles sat the child on his knee. The ensuing scene is given maximum significance by Sartre: "For the first time, I was dealing with the patriarch. He seemed forbidding and all the more venerable in that he had forgotten to adore me. He was Moses dictating the new law. My law." Sartre could be a writer, said Charles, but, recalled Sartre,

> I had to know exactly what I was in for: literature did not fill a man's belly. Did I know that famous writers had died of hunger? That others had sold themselves in order to eat? If I wanted to remain independent, I would do well to choose a second profession. Teaching gave a man leisure.[22]

The strategy worked: the little bourgeois was duly frightened by images of want. In exchange for the dream offered by his mother he substituted the more conventional one offered by Charles. Instead of becoming a great man of letters, Sartre would become a weekend-writer and school teacher like his grandfather.

Between mother and patriarch, the child had at last been given what he most desired – a future. Even so, he felt cheated. He had dreamed of winning glory as a writer; now, under Charles's plan for his future, he was destined to earn only the thin respect that would come from a few insignificant volumes. Furthermore, his writing would always be subordinated to his career as a teacher, a belief that was confirmed when he was given a "homework notebook" indistinguishable from his "novel notebook." Almost overnight the act of writing metamorphosed, for the 8-year-old, from a private pleasure into a social duty. Charles's strategy looked as if it would work: seven months passed without Sartre writing a fictional word and the old man, notes Sartre bitterly, "smiled in his beard" at the sight of his now sullen grandson.[23]

Sartre's mother's prophecy of his future destiny had been displaced by Charles's, but Sartre never forgot the original vision held out to him. He dreamed a recurring dream: near the pond in the Luxembourg Gardens, a small blonde girl, resembling his recently deceased cousin, needed his protection from an unspecified danger:

> At the age of eight, just as I was about to resign myself, I pulled myself together; in order to save that dead little girl, I launched out upon a simple and mad operation that shifted the course of my life: I palmed off on the writer the sacred powers of the hero.[24]

Making the writer a hero via the reconciliation of the desires invoked in his dream restored Sartre's interest in writing while sidestepping the choice posed by Anne Marie and Charles between becoming a professional writer or remaining an amateur one.

Like many intending writers before and since, Sartre conceived of his pen as a sword and himself as a knight-errant. He was nicely placed to effect this transformation: he merely transferred the noble attributes of the swashbuckling heroes of his "novels" to their creator. How much more gratifying that *he* should be the hero of his literary enterprise which now, after a fashion, had been validated by the patriarch. Jean-Paul, "the imaginary child, was becoming a true paladin whose exploits would be real books."[25]

He was again writing novels – or what the always plain-speaking Louise called his "lucubrations" – but he never finished them. No one read them, not even Anne Marie, and certainly not Charles, whose judgment Sartre now feared. Jean-Paul himself did not even read his own work. His act of writing was an end in itself. When he filled one notebook, he would throw it on the floor, forget it, and begin to fill another. When, for the first time in his childish life, he reread himself, the effect was devastating. He was so embarrassed by the naiveté and self-indulgence of what he had written that he stopped writing.[26]

The disillusioned writer sought consolation in his mother – and in his beloved comic books. The First World War, which otherwise seems to have affected Jean-Paul very little, stopped supplies of *Nick Carter*,

Buffalo Bill, *Texas Jack*, and *Sitting Bull*. In quest of back numbers, Sartre repeatedly dragged his mother to the Seine embankment (as he was to do with Beauvoir when they first met), where they searched the second-hand bookstalls for miles along the quays. In recalling the years of the war, Sartre stresses his intense sense of companionship with his mother. He speaks of "our union" and "our myths, our oddities of language, and our ritual jokes." His mother was cast as the friend of his own age whom he did not yet possess:

> My mother and I were the same age and were always together. *I told her everything*. More than everything: my repressed writing emerged from my mouth in the form of prattle. I would describe what I saw, what Anne Marie saw as well as I, houses, trees, people. I would assume feelings for the pleasure of telling her about them; I became a transformer of energy: the world used me to become speech.[27]

This notion of being part of a couple whose partners pool all their values and tastes and tell each other "everything" (a state of mutuality that Sartre would later call "total translucidity") is nearly identical with one of the rules that would later govern his relationship with Beauvoir.[28] The impulse to share with another, especially another woman, is shown by Sartre to be a recurring feature in the development of his character.

Midway through these years that Sartre considered his happiest, the mama's boy, now $10\frac{1}{4}$, was finally committed to a course of formal education. Charles obtained his entry to the Lycée Henri IV, a prestigious institution located in a famous building opposite the Panthéon. In his first school essay, the family pet came last in the class, but his adjustment from being the precious one-and-only to being an undistinguished one of many came more quickly and more smoothly than could have been predicted. Sartre says that the other day-students were also mama's boys, and, instead of rejecting him, as the children in the park had done, they adopted him "the very first day." He was astonished and delighted. At last he had playmates, and he made every effort to integrate himself successfully into the crowd. After school he ran, shouting, around the Place du Panthéon with the others and "played ball between the Hotel of Great Men and the statue of Jean-Jacques Rousseau."[29] For the first time in his life, Sartre felt his existence was justified.

He stopped writing "novels," and play with schoolmates replaced his daydreams of becoming a heroic writer. But the writerly dream itself, rather than being lost during his belated socialization, simply lodged deeper within him. At school, which was well stocked with studious boys, he became acquainted with peers who also "had read a great deal and wanted to write." Almost overnight, these associations transformed his wild daydreams into quiet and serious aspirations backed by social approval and realistic possibilities.

Sartre was 12 now and about to suffer the biggest jolt of his childhood – perhaps the biggest of his life. He had already acquired the rudiments of his identity as a writer and the roots of his notion of the ideal relationship. But his immediate personal resources, on the eve of his entry into a prolonged period of difficulties, were mixed. His appearance was a pure liability. Sartre was embarrassingly short, uncommonly ugly, wall-eyed, and noticeably blind in one eye. In other ways, he carried much promise but not in forms one might expect in someone subsequently identified as one of the formative minds of the century. Although Sartre had done well after settling in at the Lycée Henri IV, he did not seem to be a boy of exceptional academic promise despite his hothouse upbringing. He already possessed a remarkable facility for writing, but only in the sense that he could, with great speed and little effort, fill great quantities of paper with prose. Beyond the biased eyes of Anne Marie, the quality and content of his writing do not seem to have attracted any special regard. Though he soon did well in school, and pleased his teachers, he was not seen as exceptional. His French teacher, in 1916, comfortably noted that Sartre was "among the best in his class."[30] At this age (and for a long time to come) Sartre seems to have been more impressive as a talker than as a writer. His long years of chatty companionship with Anne Marie made him loquacious and verbally fluent to an extent that would otherwise be extraordinary in one so young. His speech also must have reflected the inner confidence of a child who had always felt much loved, because, despite Charles's and Louise's scepticism regarding some of his favourite dreams and activities, Jean-Paul seems never to have doubted either of his grandparents' devotion. And, of course, his mother's love was experienced on another plane altogether. She was, in her son's eyes – and Sartre goes to great lengths to drive this point home in his autobiography – his *friend*. To understand the older Sartre, it must be remembered that his mother was his *only* friend for the first decade and more of his life, and that throughout these formative years (which in his 50s he still thought of as "paradise") their friendship was the dominant centre of his daily existence. It was, without doubt, out of this relationship alone that Sartre formed the sensibilities and the expectations regarding friendship that he now carried into adolescence.

That adolescence began, for Sartre, with violence and betrayal. In the spring of 1917 Europe was at war and Anne Marie remarried. Jean-Paul remained with his grandparents for six months, painfully adjusting to the idea that he had lost his imperial place in his mother's life. In November, he joined the newlyweds in their comfortable house in La Rochelle, where the usurper of his mother's affections, Joseph Mancy, had been transferred recently to oversee the shipyards. Mancy, a polytechnic classmate of his stepson's father, had admired Anne Marie in his youth, but his working-class background made him an unsuitable marriage candidate for her. He was now 43; he had acquired bourgeois manners, and his managerial contempt for members of the class from which he came earned him both advancement and respect.

Charles, always a humanist, had raised his grandson to respect democracy. Mancy was neither democrat nor humanist, and, worse – in his stepson's eyes – was extremely authoritarian in his domestic setting. Charles, in appearance and manner, may have resembled God the Father, but the Schweitzer family order was a matter of compromises made by three strong adult personalities, and weighted with consideration for Jean-Paul's predilections. Under Mancy's new order, Anne Marie displayed alarming subservience to her new husband, who, far from showing due regard for her son's obvious literary propensities, decided he would be trained as a mathematician or a physicist. Sartre persisted with his own ambitions, but found himself forced into nightly geometry and algebra lessons administered by Mancy and observed by Anne Marie, who accepted her husband's well-meant advice regarding the best preparations for her boy's future. Sartre perceived the attentions of the man he saw only as a hated interloper as abusive, and, unsurprisingly, Mancy's blunt pedagogy failed with his hostile pupil. Their sessions often ended with Sartre being slapped by his mother. In his old age, Sartre could still hear those slaps: ". . . and whack! whack!, she slapped me twice across the face." Their significance hurt more than their sting: ". . . it was all over. She had been mine, totally mine. Now forced to choose, she went against me. I had become a stranger. I was no longer in my home."[31]

Sartre's psychological wound from his mother's shift of loyalty was so deep that interviews given in his late 60s show that he never came to terms with it emotionally:

> I think that one of the important aspects for me about this marriage . . . was that it forced me mentally to break with my mother. It was, if you will, as though I didn't want to be hurt, and therefore, to avoid it, I decided it would be better to make this break.[32]

This is the logic of the jilted and still unrecovered lover. "I didn't want to be hurt," Sartre says, but he, on the threshold of puberty, received the deepest wound of his life. "I had decided it would be better to make this break" – but his mother had already made the break for him. She had moved out of his bedroom into Mancy's; she had deprived him of the daily adoration of Louise and Charles; she had conceded his future to the new and more vigorous patriarch. In 1974, Beauvoir interviewed Sartre and tried to get him to talk about Anne Marie and Mancy. Sartre balked. Beauvoir prodded him with the statement that Mancy had "stolen your mother from you."

> *Sartre*:   It would take a long time to explain the nature of my relations with my stepfather.
> *de Beauvoir*:   They were the relations of childhood and adolescence.
> *Sartre*:   Yes. Let's not talk about that now, chiefly because *it hasn't the slightest importance as far as writing is concerned* . . . he knew I wrote, but he didn't give a damn. Furthermore, these pieces did not deserve that anyone should give a damn about them. But I knew my stepfather took no notice. *So he was perpetually the person I wrote against. All my life. The fact of writing was against him.*[33]

Perhaps no philosopher of note has ever contradicted themselves so completely in such a brief space as Sartre in this account of the impact of his mother's second husband on his life. The pain and dismay that Mancy caused the young Sartre was a crucial factor in his formation as an adult.

If Sartre's idyll with his mother had been shattered by her remarriage, life had also turned sour for Sartre at his new school in La Rochelle, which, unlike the elite Henri IV, was open to the public, coeducational, provincial, and, for him, very rough. Ugly and undersized, Parisian in dress and speech, and obviously thinking the world of himself, Jean-Paul was a natural victim, and, in his words, "became a whipping-boy for all the kids at the lycée."[34] Beaten up regularly, not able to fraternize freely with the other children, and made a general figure of fun, Sartre's personal hell of Others continued for at least the first two of his three years at La Rochelle. Together with his mother's "betrayal," this persecution scarred him deeply. Painful as it was at the time, Sartre's ordeal carried far greater positive effects in the formation of his character, and, ultimately, his intellect, than he ever seemed to realize. Confronted at home and at school with radical devaluation by Others, Sartre struggled from the age of 12 to 15 not to let his social identity become the basis of his self-identity. Unable to reverse others' attitudes toward him, Sartre's only means of preventing his new, inferior, public persona from taking over his inner being was to become defiant and assign his own private meaning to "Jean-Paul Sartre." At the age of 12 he was violently thrown back on his inner resources, that is, onto his own subjectivity, or what he later described as "that deep subjective reality which was beyond everything that could be said about it and which could not be classified."[35] It was difficult for the subjective resources of one so young to suffice for a psychological trial so long and so severe, but Sartre in his "Schweitzer decade" had accumulated an uncommonly large reservoir of self-esteem. After three years he would return to Paris not only with his personal dignity mostly intact, but also with the confidence of knowing that his self-respect could withstand the most extended onslaughts society could devise.

All of this is not to deny that Sartre found both his La Rochelle experience and his memory of it painful. Though he claimed it was out of consideration for his mother that he broke off his autobiography at the point of her remarriage, it seems more likely that his primary reason was to spare himself the discomfort of recalling at length those years following his expulsion from paradise. Near the end of his life, however, interviewers, especially Beauvoir, coaxed him into talking about this time in his life. Two of his misadventures, in particular, seem to have been traumatic.

Sartre's La Rochelle schoolmates teased him about not having a girl-friend. Sartre responded by fabricating a story about his precocious amorous experience. He countered his classmates' incredulity by showing them a letter he had persuaded his mother's maid to write, recalling their good times together in Paris hotel rooms. His peers seem to have been impressed until Sartre confided to one that the letter was a ruse. The truth became public knowledge, to Sartre's humiliation. The following year Sartre set his sights on a real girl, a 12-year-old who was pretty and much appreciated by his tormentors. He asked them about her and they encouraged him to approach her on the town's waterfront promenade. A few days later he saw her there, surrounded by his classmates. John Gerassi gives Sartre's account of what happened when he went up to her:

> "Finally," he said, and I sense that the story still pained him a bit, "I entered her circle. She had obviously prepared her reaction. 'Who is this bum with one eye that says shit to the other?' she asked loudly. I withdrew, as the group burst out laughing. I knew then that I was ugly. I had had such a hint after my grandfather cut my curls and my mother cried. But now I was absolutely certain: I was really ugly."[36]

If, in the midst of all his other troubles, Sartre was made to realize fully the extent of his unattractiveness, his traumas were not yet over. Another important misadventure seems to have come a few months later. His second school year in La Rochelle was nearing its end, and he still had not won acceptance from the other boys. He decided to buy himself into their favour. Each night for weeks he stole some of his mother's housekeeping money from her handbag to buy his classmates cakes from an expensive bakery. He seems to have held back a large portion of what he stole because one morning or night (Sartre told the story in several variants) Anne Marie discovered 70 francs in his coat pocket. In 1918, said Sartre, this was "an enormous sum." Sartre made up a story to explain the money to his mother: he had, he said, taken the money for a joke from a classmate named Cardino. Cardino's mother had given the money to her son, and Sartre was going to return it immediately. Anne Marie, taking no chances, said she would return the money to Cardino herself, and instructed Jean-Paul to bring him home after school.[37] At the lycée, a deal was struck whereby Cardino – the worst of the school bullies – would come for the money, and keep two-fifths of it for himself, with the remainder going back to Sartre. For his trouble, Cardino not only received the money but a lecture from Mme Mancy on how to avoid robbery. The bully found the lecture very amusing, and, as promised, split the money with the thief.

Sartre's classmate, however, went too far. He bought a huge flashlight (or, in one version, an impressive lighter) with his cash which, when it was discovered by his mother, could only be explained with the truth. The Mancys were notified, and Jean-Paul fell even deeper into disgrace at home. But the worst moment in the affair came a few days later when Charles and Louise arrived for a visit.

Although he had now been separated from Charles for almost two years, Sartre continued to think of him as his ally – now his only ally. After the uncovering of his theft, he was "ecstatic" to learn that his grandfather would arrive in La Rochelle in a few days:

> Now I would have a friend, for surely he would understand what my mother did not, that she had brought me into a city where the kids didn't want me, and therefore, where I had to pay to have them, that I stole to have a life. Charles would understand that, and defend me.[38]

He didn't. When told about his grandson's transgression Charles was more than vexed and the next day expressed his feelings in the matter when he accidentally dropped a 10-centime coin in the chemist's shop. Sartre recalled to Beauvoir his pain in relation to the subsequent events:

> It went ding. I hurried to pick it up. He stopped me and bent down himself, with his poor creaking knees, because I was no longer worthy to pick up coins from the ground.
> *Beauvoir*:   That must have wounded you somewhat. It's the kind of thing that wounds children.
> *Sartre*:   Yes, it did wound me.

Already "betrayed" by his mother, Sartre felt his grandfather too had now forsaken him. In another interview he called this episode "the second major break" in his young life.[39]

Given his physical defects and his fluency of speech, Sartre's singularity from the time of his adolescence was so pronounced that he was always bound to have difficulty in fitting into any society that demanded conformity from its members. But at La Rochelle his plight was made still worse by his social backwardness and the fact that his bourgeois classmates were at the age of maximum intolerance of any form of difference. Sartre, the most optimistic of souls, shadows his accounts of his La Rochelle years with the vocabulary of loneliness and despair: outsider, loner, solitude, violence, isolation, reflection, unhappiness, ill luck, misfortune, laughing stock, whipping-boy, hostility, humiliation. At school, his tormentors were the sons of the provincial bourgeoisie, who, by local custom, wore their father's clothes, cut down and retailored. This linked them strongly, in Sartre's mind, with Mancy, his home-based nemesis, whose high-handedness and high income epitomized the bourgeois ideal. But the "humiliations suffered at La Rochelle" seem the basis not only of Sartre's lifelong antagonism toward the French middle class, but also of his compassion for the world's downtrodden, though the latter commitment took a long time to develop.[40] In his youth, Sartre's compassion, as will be seen, was limited to the plight of individuals, an understandable position for one whose own sufferings had originated because of his separateness, and whose pain was experienced in solitude. It would not be until his mid-30s that, helped by Beauvoir, Sartre would make the moral leap and extrapolate his empathy from its grounding in his individual hardships to groups and classes of humankind.

It was also at La Rochelle, in reaction to his despair, that Sartre became habituated to a compensatory daydream that in adult life he, again with Beauvoir's assistance, brought to reality. Sartre described his recurring reverie:

> I pictured a whole little phalanstery of handsome young men – elegant, intelligent and strong – and charming girls. I too was there, and I ruled by my strength of mind and my charm. This fiction – a social one, in me who was so unsocial! – was certainly cherished by me out of revenge. For there was, in fact, a group facing me – but I wasn't its king, but its whipping-boy: it was entirely formed against me.[41]

This dream was to be realized in the form of the "family," a tight-knit circle of subordinate friends whom Sartre and Beauvoir kept round them until their deaths.

One of the few pleasures Sartre enjoyed regularly in La Rochelle – perhaps the only one – was playing the piano with his mother. After her remarriage Sartre seemed to regard her as polygamous, lamenting the fact that he "was now only a prince of the second rank."[42] He was driven to find moments of intimacy with her when he could, and the best time was after school, before Mancy, who disliked music, returned for dinner. For two years, Sartre had taken piano lessons in Paris without enthusiasm, but now that it was only at the piano that Anne Marie offered herself to him exclusively, he became enamoured of piano-playing and worked hard to improve his technique. In their big house "there was a large drawing room which no one entered except for receptions and where a grand piano sat in state," at which he often found his mother when he returned from school. She "played well," Sartre says, "sang very well" and "played difficult pieces."[43] As he improved, they played four-handed sonatas together. He also played alone, and at 70 recalled how he kept improving.

> I succeeded at last in quite difficult things, like Chopin or the Beethoven sonatas, except for the very late ones . . . And I played Schumann, Mozart, and melodies from operas or operettas, which I would sing. I had a baritone voice but I never studied singing. Nor the piano, really: I never did five-finger exercises, but by practising the same passages over and over I managed to play them in a more or less acceptable fashion. I even gave piano lessons when I was twenty-two years old, at the École normale.[44]

Sartre continued with his piano-playing until nearly the end of his life.

Removed from the bookish Schweitzer environment, without Charles's library, and with the added discouragement of his mother and stepfather reading a book only "now and then," Sartre, in his three years at La Rochelle, read only detective and adventure stories and the occasional popular novel. He credits his grandmother, Louise, with the fact that he read even this much.

> At La Rochelle I belonged to a lending library; that is to say I took over my grandmother's role. I had known about lending libraries, as I said in *The Words*, through my grandmother, who used to go to one and take out novels. So I began going to the lending libraries at La Rochelle . . .[45]

If Sartre as reader struggled to remain active, Sartre the writer really wilted when removed from the rarefied atmosphere of 1 rue le Goff. In his first year in the provinces he wrote "one last novel," and in his second and third year he wrote "much less, and perhaps not at all." His reasons were social:

> . . . at La Rochelle there was nothing that warranted my choice of writing any more. In Paris I had schoolmates who had made the same choice as

myself, but at La Rochelle there was not a single one who wanted to become a writer.

Even more importantly, he no longer received any encouragement to write at home, not even from Anne Marie. In withholding support, Sartre's mother was probably not so much pandering to the prejudices of Mancy as joining him in being a responsible parent. Her son had reached the age when he should be abandoning his childhood fantasies in favour of thinking about what he might do with the hard and limiting realities of the adult world. If her darling's writing had developed to the point of showing special promise, then the game she had encouraged him to play at a younger age might now have merited consideration as a grown-up occupation. But the truth was that, after all those feverish years of scribbling, Jean-Paul had not progressed beyond his crude recasting of children's adventure stories. At the age of 15, Sartre says, he had not yet even conceived of writing as a mode of self-expression;[46] nor is there evidence that the textual quality of his prose was advanced for his age. Worst of all, he still never redrafted what he wrote – compelling evidence that he had neither the patience, nor the tenacity, nor the ability for self-critique needed to become a professional writer.

The marriage to Mancy had saved mother and child from the uncertainty of financial dependence on a man of 80. It had also provided Jean-Paul with a stepfather and a more conventional home than he had known – both things that Anne Marie must have thought beneficial for his development. But in the first three years of her marriage, her son – to whom she was no less devoted than previously – had turned from a joy into a steady and increasing source of disappointment. His relationship with his stepfather had been a disaster from the start and had not improved. Among his provincial classmates he was a social outcast, and his "genius" had vanished without trace, leaving him, in his words, "among the middling boys, sometimes a little higher than the average, sometimes lower . . . ."[47] His theft had shocked the family, and he was profoundly unhappy, which, in the eyes of Anne Marie, may have counted most of all. Anyhow, it was decided that the 15-year-old failure would be sent to a boarding school in Paris and would spend his weekends with Charles and Louise. A new place was found for the former genius at the Lycée Henri IV, the place where, as a day-boy, he had once been so happy.

Sartre's return to Paris as a boarder at Henri IV was a positive reversal of fortune as great as the negative one he suffered with his mother's remarriage and the move to La Rochelle. The Lycée Henri IV was regarded as one of the best in France and Sartre's readmission to its privileged, all-male cloisters meant that he was now back on France's most elite educational track, and, in consequence, would for the next nine years be recipient of a hugely disproportionate share of his country's educational resources.[48] Given his subsequent ideals and his relationship with Beauvoir, it is appropriate to emphasize that the means by which he, at 15, rose to his ultra-privileged position were not academic. His place "among the middling boys" of La Rochelle did not entitle him to entry at the prestigious lycée. What secured his place at the school was the combination of his stepfather's money and his grandfather's influence. Charles used his connections to obtain a place for his unworthy grandson at the Lycée Henri IV and Mancy paid the bills. Sartre understood this well: he said "I was in Paris because my grandfather, a teacher of German, had colleagues there, headmasters who knew him and who would find me a place in a good lycée. . . ."[49] Thus, in as far as Sartre can be said to have been the product of his education, he was the creation of positive discrimination common for well-connected males – the most usual form of privilege of all.

Sartre had left Paris the most tender of 12-year-olds, but his three years of torment, besides scarring him, gave him a psychological toughness that would serve him well for the rest of his life. After the bullies of La Rochelle, he found his fellow boarders at Henri IV a gentle and sophisticated lot, and, within a few months, he was again on his social and educational feet. He even began writing again, and, under the influence of his new peers and his septuagenarian grandmother, "began to read serious things."[50] Weekends and Thursday afternoons he spent at his grandparents' flat, and now, after his experiences in the world, it was not Charles but Louise in whom he found a kindred spirit. "Alert as ever" and keen to deepen her grandson's mind, she reread and thereby encouraged Sartre to read Dostoevsky, Tolstoy and Stendhal. She engaged him in long conversations about their works. Recalling these tutorials, Sartre says that, when discussing the Russian revolution with her, "I was stunned to hear her say that the poor had no choice but to use violence to better their lives." It would be a long time before Sartre would become interested in politics, but he took to Dostoevsky and Stendhal immediately. With the latter, the effect was so profound that Sartre's ambition became, as he announced to Beauvoir eight years later, "to be Spinoza and Stendhal, both at the same time."[51]

At the lycée, Sartre was forming his first intimate friendship since the loss of his mother to Mancy. Paul Nizan, an engineer's son, a brilliant student, and a future novelist, also wanted to become a writer, and so was a natural ally for Sartre. During the latter's previous stint at Lycée Henri IV, the two had been casual friends; now, as boarders, they became constant companions, slept next to each other in the dormitory, and read each other's stories. The new boy from the provinces found his fellow aspirant the better writer and also, like most of his classmates, better read. As he was later to do with Beauvoir, Sartre began competing with his friend, striving, particularly in his reading, to catch up. Besides carrying out the programme laid down by Louise, Sartre devoured the works of the twentieth-century writers favoured by his fellow boarders: Gide, Giraudoux, Valéry, Conrad, Jules Romains, Paul Morand, and, especially, Proust. He also began to excel in the classroom, and, in the security of knowing that his fellow students were modestly intimidated by the rough ways he had acquired at the hands of his port-town tormentors, he became extroverted – known and appreciated for his wit and antics, as well as for performances as a jazz singer. He was chosen as the organizer of celebrations. For the long-suffering mama's boy this new popularity must have seemed too good to be true, the way into a new kind of paradise. At the end of the year he passed the first half of the *baccalauréat* and won the prize for the best student in his class.

In France, students in their final year of secondary education are required to study philosophy, and at the Lycée Henri IV Sartre became convinced that, as a future novelist, philosophy was important for him. "I thought that if I specialized in philosophy I would learn the entirety of the world that I was to talk about in books. It gave me the raw material, you might say."[52] As a means for learning about the world, Sartre's choice of philosophy at the age of 16 was retrogressive when compared to his earlier preference for the encyclopedia in his grandfather's study. But his heightened idealism was offset by the immersion in the material excitements of Paris that he was sharing with Nizan during their long walks through the capital. His friend, as Sartre explained, "was much more curious about the outside world than I was."[53] Left to his own devices, Sartre's curiosity about the tangible world, unlike Nizan's and, later, Beauvoir's, would never be very great. His decade of house-arrest, six stories up, imbued him with a permanent indifference regarding terrestrial exploration. A friend, however, could set him going, and, once in motion, he usually enjoyed himself as well as being a good companion. Just as he relished the memory of his earlier jaunts along the Seine with Anne Marie, he rapturously recalled his teenage perambulations with Nizan.

> We walked around Paris for hours, for days. We discovered flora, and fauna, stones, and we were moved to tears when the first neon advertisements were switched on. We thought the world was new because we were in the world . . . One night these supermen at large climbed the hill of Sacre Coeur, and saw at their feet a jeweller's shop in disarray. Sticking his cigarette into the left corner of his mouth and grimacing hideously, Nizan said "Hey, hey, Rastignac!" I repeated "Hey, hey!" as I was intended to, and we walked down, pleased with this discreet revelation of our literary knowledge and the extent of our ambitions.[54]

But "ambition" is not quite the right word in Sartre's case. The year in which he looked down on the lights of Paris with his friend, his childhood wish for a life governed by destiny reemerged as a religion of personal greatness. Surrounded day and night by highly privileged male adolescents, the restraint Charles had imposed upon his grandson's dream was stripped away. Sartre decided – this time unshakeably – that he was destined to be a *great* writer. Furthermore, a great writer, according to his faith, had "a great man's life," which Sartre vividly imagined:

> . . . there were solitude and despair, passions, great undertakings, a long period of painful obscurity (though I slyly shortened it in my dreams, in order not to be too old when it ended), and then glory, with its retinue of ambition and love.[55]

Sartre says he had no evidence to support his belief in his destiny, but "believed in it as a Christian believes in the Virgin." At La Rochelle, waiting for a bus, he had decided he did not believe in God; he filled the vacuum with his exaggerated belief in himself. His early years in his grandfather's study had planted in him the romantic's ideal of the great man who wins immortality for himself through his art. At 16 he took the lives of Shelley, Byron, and Wagner "as models." He recalled how he

> was extremely conscious of being the young Sartre, in the same way that people speak of the young Berlioz or the young Goethe. And from time to time I'd go for a little stroll into the future, for the sole pleasure of looking back from up there at my young present and shaking my head – as I believed I should do – saying: "I never thought suffering would be so useful to me, etc.": as an old man . . .

Others had the Catholic or the communist faith to sustain them. Sartre, at 16, looking down on Paris from the Sacre Coeur, had the faith that his life was a great "undertaking favoured by the gods."[56]

The "great life" Sartre saw before him called for him to attend the Ecole Normale Supérieure (ENS), an institution of such exclusivity as to be without parallel in the English-speaking world where neither Oxford nor Cambridge nor Harvard can compare in its selectivity. It admitted only 25 men a year to study humanities and about the same number in the sciences, giving a total enrolment of nearly 200, all of whom were men who "were considered and thought themselves to be the chosen of the chosen."[57] Any chance of being placed in the top 25 of the Ecole's humanities entrance exam meant cramming for two years or longer at one of the elite schools dedicated to this very specialized educational task. The Lycée Henri IV was among them, but around the corner was the Lycée Louis-le-Grand, whose success rate at getting its boys into the ENS was unrivalled. In the autumn of 1922, after passing the second half of his *baccalauréat*, the 17-year-old Sartre and his friend Nizan moved to Louis-le-Grand.

At the new school, Sartre and Nizan became close friends with René Maheu – fellow student, future director of UNESCO from 1961 to 1974, and Beauvoir's first lover.[58] Sartre, who at about this time lost his own virginity in a single encounter with a doctor's wife, continued to write, and, along with Nizan, published stories in a student journal. Of the three friends, Sartre was the weakest student and Nizan the strongest. Their curriculum was divided into the five disciplines that comprised the ENS entrance exam: Latin, Greek, French, History, and Philosophy. Sartre's marks placed him near the bottom in French, near the top in none, and, overall, in the middle of the second quartile.[59] The fact that Mancy and Anne Marie had moved back to Paris and had insisted that Sartre live with them may have influenced his performance, which was surprisingly mediocre after his two years of brilliance at the Lycée Henri IV. But this lackluster performance was nothing when compared to the academic humiliation that awaited him in university.

At 17, Sartre disdained fashion in favour of academic slovenliness, whereas Nizan, who continued as his best friend, was an elegant dandy who followed fashions closely, carried a walking stick, and affected a monocle.[60] This polar opposition in style of dress mirrored an equally fundamental difference in their moral temperaments of which they were highly aware and which placed an increasing strain on their relationship as they grew older. Sartre's troubles with his stepfather and La Rochelle schoolmates made him permanently hostile to hierarchical orders, and especially to any form of paternalism. Furthermore, it was in his rebellion against those he felt to be his oppressors that he discovered what he perceived as his freedom to choose his being. When he returned to Paris, he generalized from his experience and claimed subjective freedom for all. Sartre put the case for this egalitarianism to his new friend. But Nizan adored and was fervently loyal to his managerial father, which strongly predisposed him to paternalism, especially in its more autocratic forms. In his teens, he wavered between entering the Catholic priesthood and becoming a Communist Party worker. When Sartre tried to convince him of his metaphysic of freedom, Nizan countered with his more fashionable one of determinism. Neither youth was swayed, but for both, these were probably their first important philosophical dialogues. Their subsequent careers confirmed their late-adolescent positions: Nizan joined the communists, and Sartre's books were banned by both the Vatican and the Kremlin, to his evident and anti-authoritarian satisfaction.

* * *

In October 1924, Sartre's destiny unfolded according to plan, taking him a few streets south of the Panthéon to the ultra-exclusivity of the Ecole Normale Supérieure. Nizan went too, and, again, they slept in adjacent beds in the dormitory and shared a study. The following year Maheu joined them. They were all studying philosophy. In four years they would sit the *agrégation*, which, if they passed, would guarantee them employment as teachers. Sartre had opted decisively for philosophy after reading Bergson's *Time and Free Will*, a major step in his formation as a philosopher as it showed him the possibility of connecting philosophy with psychology and thus to personal experience. Such a philosophical programme dovetailed not only with his intention to become a novelist, but also with his predilection to consider the position of the individual relative to society. Upon arriving at the ENS Sartre also placed himself "under the sign of Descartes," and this unlikely combination of Cartesianism and Bergsonianism would prove highly fertile. The former's scepticism and rigorous analysis provided a hard centre for the latter's intuitionism.[61]

In his second year at university, Sartre befriended Pierre Guille, a sardonic and suave literature student who joined Nizan and Maheu to form the nucleus of Sartre's band of associates. When Sartre recalls how the four of them "formed a group that people made fun of" and that he had his nose bloodied "from time to time," one is reminded of his fate at La Rochelle. But there Sartre had been solitary and an unwilling outcast, desperate to win acceptance. Now he had friends with whom he openly invited the hostility of other students by showing contempt for them. Usually Sartre's taunts were verbal, but sometimes the friends resorted to more juvenile pranks, which Sartre saw as carrying serious meanings:

> I believe that when we hid at the top of the stairs in order to throw water bombs at the boys who came back about midnight wearing dinner jackets, having been dining out, we were thereby indicating that dining out, the dinner jacket, the distinguished air, the well-brushed hair were wholly exterior things, nonvalues, of no worth. That these were things those boys ought not to have possessed, ought not to have desired, because what one ought to desire was the inner brilliance of genius, and certainly not success at a fashionable dinner party.[62]

Sartre's small band of rebels, defining themselves in opposition to bourgeois values symbolized by the dinner jackets, was a dress rehearsal for the "family" of friends he would create with Beauvoir. Sartre was now, as he would remain for the rest of his life, an "outcast" by choice.

Sartre's first serious love affair began when he was 19. Attending a paternal cousin's funeral in his father's native Thivers, he caught the eye of a beautiful young woman in black, befriended her at the family dinner that followed, and spent that and the next three nights with her. She was 22 and still living with her parents in Toulouse, but was an experienced sexual eccentric who frequented fancy-dress orgies and dabbled in prostitution. Encouraged by Sartre, she would soon come to Paris, eventually to make her name as an actress and playwright under her stage name of Simone Jollivet. Later she would help Sartre break into print as a novelist and provide him with an entrée into the Paris theatrical world. Their affair, which, in the main, took place by letter, lasted two years, and "Toulouse" and Sartre remained friends until her death in 1967. Beauvoir met her in 1929 and described her as "a beautiful woman, with vastly long blonde hair, blue eyes, a delicate complexion, an alluring figure, and perfect wrists and ankles." Concerning Jollivet's working nights in Toulouse, Beauvoir illustrated her flair for pandering to male sexual fantasies.

> She would say goodnight to her mother (of whom she was very fond) with great tenderness, make a pretence of going to bed, and then slip out . . . she possessed an acute sense of the appropriate *mise en scène*; while awaiting a client in the room set apart for him, she would stand in front of the fireplace stark naked, her long hair combed out, reading Michelet or, at a later period, Nietzsche. Her cultured mind, her proud bearing, and the subtle technique she brought to her task knocked down town clerks and lawyers flat: they wept on her pillow from sheer admiration.[63]

Sartre probably wept too: he said his first affair "was an unhappy one" – especially after he discovered that Jollivet was conducting four more affairs simultaneously. He wrote a melancholy novel about his experience, *Une Défaite*.[64] But in another and more lasting sense, the affair was a great victory for Sartre, as it showed that by the force of his charm and personality he could overcome his physical shortcomings to win the love of beautiful women.

Sartre also sought to overcome his physical handicaps in the gym, where he learned to box. He became so muscular through the back and shoulders that, as his college friend Raymond Aron recalled, "he climbed up ropes, legs at left-angles, with an effortless ease that stupefied us."[65] Aron, son of a wealthy Jewish family, was also destined for writerly fame, and, beginning at the ENS, became one of Sartre's lifelong intellectual rivals. In their third year Nizan went abroad to work as a private tutor in Aden, and Aron replaced him in Sartre's study cubicle where they engaged in endless discussions. It was also in 1926 that Maurice Merleau-Ponty, future existentialist philosopher and long-term colleague of Sartre and Beauvoir, was one of the lucky twenty-five humanities students admitted to the

ENS. In addition to Nizan, Aron, and Merleau-Ponty, another half-dozen of Sartre's contemporaries at the Ecole would become leading actors in French cultural life. These and other *normaliens* comprised one of the most powerful networks of influence in French social and political institutions. For example, when, in 1925, in the students' annual revue, "Sartre danced in the nude with a semi-nude Nizan," *normaliens* in the audience included the Prime Minister and the Head of the Chamber of Deputies.⁶⁶ The following year Sartre's stage performance was lauded in the national press, and, in 1927, he used the revue to launch a vitriolic attack on French jingoism and war profiteering. A few weeks later, Lindbergh made his historic flight across the Atlantic, and Sartre informed the newspapers that at 9.30 a.m. the aviator would appear at the ENS. Thousands, including 500 journalists, congregated outside the Ecole's gates and cheered as a student who resembled the hero arrived and was carried aloft by Sartre and friends. When the truth emerged, a scandal ensued, and the ENS's famous director resigned.

However, for all his seeming rebelliousness, outrageousness and protests, there was another, and almost secret, side to Sartre which respected bourgeois conventions. In the summer of 1927, while spending his vacation at the home of a fellow *normalien*, he met, wooed, and became "unofficially" engaged to his host's cousin, whose father was a wealthy Lyon grocer. This arrangement may have been inspired by Nizan, who, returning from Arabia, joined the Communist Party and married a banker's daughter. Little is known about Sartre's fiancée, although the relationship lasted a year, during which he, as in the previous year, devoted himself mainly to preparing for the *agrégation*. In July, to everyone's astonishment, Sartre failed his written exams. When he travelled with his mother and stepfather to Lyon to ask his fiancée's parents for permission to marry their daughter, news of his failure had preceded him and his request was turned down. Sartre was crushed: "I took a bottle and went off into a field, alone, and there I drank, and even wept."⁶⁷

Like most students who fail their examinations, Sartre prepared a classic excuse for his bad results. The explanation has been used throughout the centuries: he was too original in his answers. By this logic, his close colleague and debating partner, Aron, must have been the least original of all those sitting the exams because he had been placed first. Sartre promised himself he would do better when given a second try at the *agrégation* the following summer. Meanwhile, in disgrace, he was forced to leave the Ecole Normale Supérieure and moved into a dormitory for ordinary university students. He was 23, and his dream of becoming a great man and leading a great man's life must now have weighed heavily on his spirit.

2
THE PHENOMENON WHO
COULD NOT BE JUDGED

In the first paragraph of *Memoirs of a Dutiful Daughter* (1958), the initial volume of her autobiography, Simone de Beauvoir emphasizes her early sense of self-worth and superiority. Reflecting on her feelings at the time of the birth of her sister, Beauvoir stresses her own primacy: "As far back as I can remember, I was always proud of being the elder, of being first."[1] This privileging of the self, of claiming ascendancy, of taking the first place as a matter of right and identity is one of many signs in Beauvoir's account of her rebellion against the notion of the woman she was supposed to become. The role of *jeune fille rangée* was precisely what she rejected. Those who posit a Beauvoir who was a strictly relative creature, if not of the milieu of her birth, then of a later capitulation to Sartre, which mimicked the usual forms of female destiny as subsidiary, reflective, derivative, mistake both Beauvoir's anarchic rejection of conventional patterns in either their positive *or* negative forms, and the daring of her continual confirmation of her right to invent for herself standards of her own choosing.[2]

The geographical circumstances of Beauvoir's birth were ludicrously prescient. Simone-Ernestine-Lucie-Marie Bertrand de Beauvoir was born on January 9, 1908, in a flat on the Boulevard du Montparnasse whose windows overlooked the Boulevard Raspail just above what was soon to become the café La Rotonde.[3] Her parents were members of the bourgeoisie whose social place and ambitions were gradually effaced and eroded. Like Sartre, Beauvoir spent most of her adult years within the radius of a few miles of her birthplace, living much of her intellectual life precisely in the public gaze of the patrons of those Left Bank cafés so closely associated with her and Sartre's later careers. The mores of her family were quite other. Though her parents were not well-off, Beauvoir was brought up in the air of exclusivity and snobbery appropriate to a daughter of the bourgeoisie of the Belle Epoque, an atmosphere made all the more important because of her parents' inability to maintain the style of life to which they more than aspired. What they took as their social birthright never quite crystallized into fact.

Unlike Sartre, who was born into much the same social stratum, Beauvoir's upbringing as a female of her time and class was simultaneously privileged and deprived. While Sartre was confronted early with the idea of literary genius, which he quickly annexed to his already princely notion of himself, aligning himself with the noble men who ranged from his grandfather near at hand, to the more distant (but still attainable for a "genius") greatness of Victor Hugo, Shakespeare, and the other male denizens of his grandfather's classic library, Beauvoir was schooled by her mother in the attributes of a hyper-respectable Catholicism considered fitting for the future wives of her class. Examples of female superiority, in any area outside the family, were absent during Beauvoir's early years. The single semi-exception was a woman Beauvoir recalled as an "obese and bewhiskered aunt who wielded a pious pen" who used one of Beauvoir's frequent temper tantrums as an illustration for the edification of other young miscreants in a children's magazine chillingly entitled *La Poupée modèle*.[4] While there is a certain irony in Beauvoir already being immortalized for her rebellious sins at the age of $3\frac{1}{2}$, the example provided by Great-Aunt Titite was scarcely encouraging.

However, like Sartre, and like other beloved children, Beauvoir's childhood also provided continuing affirmations of her worth. She recalls herself as a centre of importance within the family and remembers the attention she received as the first child in terms that were clearly important for her development:

> At home, the slightest incident became the subject of vast discussions; my stories were listened to with lavish attention, and my witticisms were widely circulated. Grandparents, uncles, aunts, cousins, and a host of other relatives guaranteed my continuing importance. In addition, a whole race of supernatural beings were forever bent over me, I was given to understand, in attitudes of divine solicitude . . . My heaven was constellated with a myriad benevolent eyes.[5]

Looking back on her childhood, Beauvoir remembers herself as "sheltered, petted, and constantly entertained by the endless novelty of life," yet aware, obscurely, that even her child self was attuned to an underlying dissatisfaction that expressed itself in frequent blind rages which her parents failed to control, thus allowing her to develop in directions that a better-disciplined little girl might have found difficult. Beauvoir's first form of rebellion was that of pure emotion, and it worked distinctly to her advantage. In *Memoirs of a Dutiful Daughter* Beauvoir notes that, in her late 40s, she still wondered about the sources of her terrific displays of refusal in her childhood and offers as a partial and tentative explanation her unwillingness "to submit to that intangible force, words" when applied as unexplained interdiction, which she, as a child, interpreted as the rule of force rather than necessity. "At the root of these implacable laws that lay as heavily as lead upon my spirit I glimpsed a sickening void: this was the pit I used to plunge into, my whole being racked with screams of rage." The void of conformity to the rules of her class and gender, and, indeed, her age, could be escaped through personal refusal, or, at least, this was the lesson Beauvoir drew from the success of her anger. And while Beauvoir says that, as a child, she "accepted without question the values and tenets of those around me," that she "never seriously called authority in question," her early habit of revolt against the imposition of adult power over the weak child formed the bedrock beneath her adult rebellion.[6] That her tantrums worked, that they, indeed, delighted her parents with her strength of character, and were, at times, reported to others with pride, also allowed for the development for optimism in revolt. Beauvoir also linked rebellion with joy. Few have recorded as absolute a desire for happiness as Beauvoir; it remained one of her chief demands from life. The Beauvoir of the future was in the making through a slight mutation of bourgeois self-confidence encouraged in a little girl who was to take her prerogatives for self-determination further than her amused parents could possibly have foreseen.

Those parents embodied the dichotomy between secular rationalism and religious dogmatism through which Beauvoir herself was later to navigate. But this paradigm, which is often the interpretation given (even by Beauvoir herself) to the force of the conflicting ideologies represented by her humanist father and Roman Catholic mother as a perfect reflection of typical gender splits of the time is, in fact, not so clear-cut. Beauvoir's parents were as fundamentally similar as they were divided. Both had a keen sense of class identity; both valued literature; both respected intelligence and education; and both, importantly, encouraged Beauvoir in her childish attempts at writing.

Commentators are agreed that the marriage, on December 21, 1906, of Georges Bertrand de Beauvoir, second son of a landowning family in the Limousin area of the south of France, and Françoise Brasseur, eldest daughter of a Verdun banker who was to go bankrupt before being able to pay her dowry, was initially based on love and mutual esteem, even though contracted through the "powerful marriage-broking networks" that then were still in place for the children of the *haute bourgeoisie*.[7]

The marriage, in time, fell into typical patterns of disappointment. Georges de Beauvoir was a younger son in a family rich enough to give him a taste for the life of an aristocratic aesthete, but not enough money to secure it. He made his living, grudgingly, as a lawyer, but would have preferred to be an actor, had this not violated his class prejudices. Georges was a dandy, to whom the cautious financial rigors of a respectable marriage did not come easily. Remembering her father in 1985, Simone de Beauvoir recalled both his attitudes and his sense of personal defeat: "He always made a scene when asked for money . . . One day he told a friend that he was a failure. He despised work, thrift, a dutiful life because he thought of himself as an aristocrat, an artist and all that."[8] Although Georges de Beauvoir continued to work at a variety of ever less distinguished posts during his lifetime, his energy was saved for the amateur theatricals which were his passion, and for his public life in the cafés and drawing rooms of Paris. Little of this included his family. Georges de Beauvoir's political views were right-wing. He was an anti-Dreyfusard, a supporter of Charles Maurras's *L'Action Française*, and fostered a long-term interest in the racist theories of Gobineau, which he took care to relay to his wife and daughters.[9] His interest in literature, however, was equally constant, and something that he also brought to the family's life. And through her father, Simone de Beauvoir was given the two-and-a-half months of each summer spent at her paternal grandfather's estate at Meyrignac and her Aunt Hélène's Château de la Grillère close by, which seemed to her full of treasures and magic, and which were the initiation into her of love for the countryside which prompted the extensive, and sometimes solitary, walking tours of her adult life.

Françoise de Beauvoir took her marriage seriously, so seriously that the strain of behaving properly in all things wifely and motherly turned into a rigid manner whose sole purpose seemed focused on propriety and control. Beauvoir's earliest memories of her mother were "of a laughing, lively young woman." But to the family maid, Louise Sarmadira, and to her daughters "she showed herself to be dictatorial and overbearing." Aware of her provinciality, Françoise clung to her sense of duty in marriage for an identity. She found her convent morality, which she valued, useless in Georges's Parisian circles. As a substitute, she dedicated herself to etiquette and respectability. Herself the most formidable of formidable women, Beauvoir remembered her sense of her mother's power; as a child, Beauvoir "made no distinction between her all-seeing wisdom and the eye of God himself." While gladly receiving the literary education that her husband thought it incumbent to give to his wife ("'The wife is what the husband makes of her: it's up to him to make her someone,' he often said"), Françoise's world narrowed further as the family fortunes declined and was not to open up again until after the death of her husband, by which time her influence on her daughters had become marginal.[10]

In 1919, after Georges moved his family to a cheaper apartment at 71 rue de Rennes (a great loss to Simone who could no longer watch the people passing in the street from a balcony), Françoise's social contact became restricted to the extended family. With the Beauvoirs' reduced circumstances (her dowry forever unpaid, Georges's prospects ever fading), Françoise could neither afford to entertain nor to dress herself or her children to the expensive standard of the social set to which she aspired. She, to her shame, made do without that shibboleth of respectability, a live-in servant (Louise, the mainstay of Simone's early childhood, married, lived in a wretched attic, and suffered through the death of her baby which gave Simone her first frightened introduction to mortality), and became increasingly lonely. Françoise, an intelligent woman, who had once dreamed of becoming a teaching nun, turned her considerable energy to duty, to tight household management, and to the close control and education of her daughters.[11]

Simone de Beauvoir's tendency to think of herself as (in her repeated phrase) "the One and Only"[12] was encouraged by her intellectual precocity. Like Sartre, she learned to read early, before she was 4. She loved books, was from the first a prodigious reader, and was not quite sure whether, when she was an adult, she would like to write books or to sell them, but, either way, she felt that books "were the most precious things in the world," that literature was the only area in which she could venture her "creative talent," and that, while her attention to objects was cursory, she did "know how to use language, and as it expressed the essence of things, it illuminated them" for her.[13] While Beauvoir, in her 20s, was to lose her faith in the transparency of language to the material world, it was an idea that served her well in the first stages of her life.

Both of Beauvoir's parents encouraged their daughter's literary interests. Both read and admired her early attempts at writing stories, and Beauvoir took seriously her father's idea that "there was nothing in the world finer than to be an author."[14] Georges's opinions on matters artistic and intellectual were important for the young Beauvoir, and she responded readily to the directions given by the father she adored. Indeed, this acceptance of intellectual influence was her closest connection with her father. As she recalled it, to Georges, she "was neither body nor soul, but simply a mind." Her mother, entrusted completely with her daughters' "bodily and moral welfare,"[15] also eagerly committed herself to supervising her daughter's intellectual development, and managed, somehow, to pay for what she considered the best education for her children. She learned Latin and English to help Simone with her studies. A woman who had loved to learn in her own childhood, Françoise helped with her daughter's homework, encouraged growth in intellectual directions which led to conclusions she could not have foreseen, while at the same time imposing on Beauvoir notions regarding the danger of initiative, fear of and contempt for the body, the inescapability of the laws of conventionality, and the necessity for piety. In the end, like so many mothers who wish to live through their offspring, Françoise was perceived by her daughters as a resented maternal presence who censored their reading material, and who did not stop reading their mail until Simone was almost 20.[16]

In comparison to Sartre, Beauvoir's early formal education was ludicrously inferior. In October 1913, at the age of 5½, she was enroled in the Cours Adeline Désir, a compromise choice for her parents who could not afford the convent education Françoise would have preferred but which avoided the humiliation of a state school which Georges thought beneath the family dignity.[17] Le Cours Désir was a highly respectable Catholic girls' school in which the mothers of the pupils were expected to attend classes as often as possible. The teachers were members of a lay order affiliated with the Jesuits, who provided an education which was rigorous but timid, relying heavily on memorization, religious instruction, and outdated material. A major aim of the school was to give the girls an ever more complete knowledge of Thomas à Kempis's *Imitation of Christ*. The institution was officially considered only a primary school, and even after attendance of ten to twelve years the students' education did not quite reach the standard of O levels in England or much more than that of junior high school in the United States.[18] Beauvoir, with her mother's constant support and encouragement, threw herself into her schoolwork with pleasure and continuing success. She would later concur with her father's low opinion of the teachers at Le Cours Désir; he made "no secret of the fact that he found these pious old frauds a little backward."[19] In her adolescence, Beauvoir understood only too well the inadequacy of her education when compared with that of boys of her age and background:

> Papa used to say with pride: "Simone has a man's brain; she thinks like a man; she *is* a man." And yet everyone treated me like a girl. . . . I was confined to the nursery. But I did not give up all hope. I had confidence in my future. Women, by the exercise of talent or knowledge, had carved out a place for themselves in the universe of men. But I felt impatient of the delays I had to endure. Whenever I happened to pass by the Collège Stanislas my heart would sink; I tried to imagine the mystery that was being celebrated behind those walls, in a classroom full of boys, and I would feel like an outcast. They had as teachers brilliantly clever men who imparted knowledge to them with all its pristine glory intact. My old school-marms only gave me an expurgated, insipid, faded version. I was being crammed with an ersatz concoction; and I felt I was imprisoned in a cage.[20]

The educational distance Beauvoir had to cover to achieve her famously brilliant results in the *agrégation* at an exceptionally early age should not be forgotten. Further, the fact that she was raised in a milieu which identified intelligence, even thought itself, as masculine characteristics underscores just how difficult it must have been for this woman, who was to become one of the leading intellectuals of the century, to lay the foundations of knowledge in childhood.

Because of the family's elevated sense of its status, and the tendencies toward exclusivity which Beauvoir absorbed, other children were a rarity in the landscape of Simone's childhood. Forbidden to associate with random children in the Parisian childhood utopia, the Luxembourg Gardens (and it is curious to think that Sartre, a lonely little boy, was also playing, isolated, in the Luxembourg at the same time that the Beauvoir girls, isolated from playmates not vetted by their mother, were given their outings), Beauvoir's close acquaintance with her peers was limited. The exception, for which she was always grateful, was her sister, Hélène, two-and-a-half years younger than she, for whom Beauvoir always retained great affection. Together they invented a secret language, and together they resisted, if only by ridicule, some of the conventionalities of the family's dictates. Hélène, recalled Beauvoir, rescued her from silence and provided a confederate with whom to act out her dreams. But their relationship was also one of radical inequality. What Beauvoir said she appreciated most in her sister was the "real hold over her" that she maintained; Hélène gave Simone authority, and "one of the most durable bonds" between them was that of "master and pupil." The role Beauvoir most cultivated in relation to her sister was that of teacher. In a statement deeply indebted to the Hegelian interpretation of the master–slave relationship, Beauvoir records that, from the age of 6, "Thanks to my sister I was asserting my right to personal freedom; she was my accomplice, my subject, my creature." The bond enthralled her and at an early age Beauvoir decided that she "wouldn't have any children; the important thing for me was to be able to form minds and mould characters: I shall be a teacher, I thought."[21] Husbands, too, were encumbrances, blocking the light, creating distractions and work for their mates. She would do without.

From whatever voids of silence Hélène rescued Simone, and whatever the affection that existed and continued to exist between them, their relationship remained unequal. For equality, like many bookish and intelligent children, Beauvoir first turned to fictional characters. Although Beauvoir's childhood reading was extensive, in *Memoirs of a Dutiful Daughter* she singles out two novels as crucial in terms of character identification and interpretation of the world: George Eliot's *The Mill on the Floss* and Louisa May Alcott's *Little Women*. That these novels should have made such an impact on Beauvoir is not surprising; they are key, even classic texts for intelligent girls. For Beauvoir, Jo in *Little Women* was striking, superior to her sisters, by virtue of her "passion for knowledge and the vigour of her thought" while her sisters were only distinguishable for their "personal morality." Jo would have "an unusual life" and Beauvoir intensely identified with her.[22] Later, between the ages of 11 and 12, *The Mill on the Floss* was to supersede *Little Women* in its influence. For Beauvoir, "Maggie Tulliver, like myself, was torn between others and herself: I recognized myself in her. She too was dark, loved nature, and books and life, and was too headstrong to be able to observe the conventions of her respectable surroundings." Maggie's friendship with the hunchback Philip Wakem, based on camaraderie and the discussion of books, seemed perfect to Beauvoir; she could not understand Maggie's physical passion for the handsome Stephen Guest as a more urgent force. In Beauvoir's view, "the exchange and discussion of books between a boy and a girl linked them forever." But despite her incomprehension of the irrational power of sexuality, Maggie's story moved her profoundly:

> I wept over her story fate for hours . . . I resembled her, and henceforth I saw my isolation not as a proof of infamy but as a sign of uniqueness . . . Through the heroine, I identified myself with the author: one day other adolescents would bathe with their tears a novel in which I would tell my own sad story.[23]

The loop back to George Eliot, the leading female intellectual of the nineteenth century, who ordered her own life the way she chose in the teeth of social stigmas far more daunting than anything Simone de Beauvoir would face, is entirely apt. The formation of Beauvoir's own views of what constituted the most profound connection between a man and a woman leads directly (though, of course, with convenient hindsight) to Sartre.

It must never be forgotten that Beauvoir's most formative childhood years took place under the shadow of the general European catastrophe of the First World War and its deadly aftermath of high fatalities from influenza. Her father, who had been discharged from the army reserves because of heart trouble in 1913, was called up, along with one of Beauvoir's uncles (who was to die in the post-war epidemic). By October of 1914, Georges was at the front with the Zouaves, but within three months had suffered a heart attack, after which he was sent to a military hospital from which he was released early in 1915. He was then sent back to Paris for the duration of the war to work in the offices of the Ministry of War. Daily life became increasingly difficult as the fighting ground on, and, by the last year of the conflict, Beauvoir's sleep was broken with nightmares about refugees and the wounded.[24] Beauvoir recalled 1918 in particular, the year in which she was finally old enough to comprehend the menace that the war presented, as a turning point for her. She understood, sickeningly, that the world was not a safe place:

> It was my gradually developing powers of imagination that made the world a darker place. Through books, communiqués, and the conversations I heard, the full horror of the war was becoming clear to me: the cold, the mud, the terror, the blood, the pain, the agonies of death. We had lost friends and cousins at the front. Despite promises of heaven, I used to choke with dread whenever I thought of mortal death which separates forever all those who love one another.[25]

The war also increased the growing personal tensions in Beauvoir's family: everyone was overwrought under the strain of the war and money was ever tighter; food and coal were scarce, bombing raids were frequent. And over everything hung the threat of death which was to be one Beauvoir's deepest obsessions; not, after all, surprising in an individual whose mind was initially formed during one of the great periods of slaughter of this century.[26]

Indeed, for Beauvoir, her loss of faith in God (prepared for, even made safe, by the example of her father's religious scepticism), was marked by her realization that she was alone in the universe, "without a witness, without anyone to speak to, without refuge." This revelation was linked in her mind, as, indeed, it is in the text of *Memoirs of a Dutiful Daughter*, with a deepening of her sense of mortality, as, one afternoon, alone in her parents' flat, she understood that she would die, that she was "condemned to death." This awakening was profoundly disturbing to her: "I did not attempt to control my despair: I screamed and tore at the red carpet . . . It seemed to me impossible that I could live all through life with such horror gnawing at my heart."[27] For Beauvoir, this bleak convergence of understanding and emotion meant her final propulsion from the nineteenth-century certainties that shaped her parents' ideology, into the alienated modern condition. For the rest of her life she would circle the questions of mortality and cosmic estrangement in the midst of the everyday reality of ordinary life. This daughter of the Roman Catholic Church, this child of the First World War, the future author of such meditations on death as *The Blood of Others*, *The Useless Mouths*, and two books, *A Very Easy Death* and *Adieux*, concerned with the deaths of those closest to her, faced life unquestionably, and, in the full metaphysical sense, alone. Her attention turned irretrievably from the pieties of family, class, nation, and religion to the flesh-and-blood others who would now be the sole source of satisfaction or pain.

The complex move from what Edward Said calls "filiation," the situation inherited through birth, history, and family, to "affiliation," the conscious choice of groups, ideologies, and projects with which individuals align themselves, was difficult and complicated for Beauvoir. She herself analyzed part of her move towards affiliation as "classic" in its Freudian dynamics as she moved her allegiance from her mother, who had structured the world of her childhood, to her father, who represented for her intelligence, knowledge, and the world. It was a traditional gendered split, but Beauvoir interpreted her growing attention to her father as a matter of mental rather than physical passion. As she said, "in reality it was truly a love affair of the head."[28] She became aware that her "father's individualism and pagan ethical standards were in complete contrast to the rigidly moral conventionalism" of her mother. This divide, painful as it was for her, was also enormously productive. Beauvoir judged that this "imbalance, which made my life a kind of endless disputation, is the main reason why I became an intellectual." Whatever the internal dialogue set working in her between her mother and her father's principles, the adolescent Beauvoir craved, more than anything, the approval of her father, which was increasingly withheld as he became disappointed in Simone's lack of elegance and beauty. There was a further severance after an incident during which Georges declared vehemently that no child has the right to criticize their mother, even if the criticism is true. The impact of this statement, which represented yet another iron law, another bar to the cage from which she wished to escape, from the parent who was identified in Beauvoir's mind as the partisan of the truth, was deep. Stung by Georges's remarks, which were made about a cousin, but which corresponded to her own increasingly critical views of her mother, Beauvoir chose truth over loyalty. She no longer believed in her father's "absolute infallibility." Still too young to leave the family home, she nevertheless knew that she would have to keep parts of herself hidden, especially from her mother, who demanded complete, confessional disclosures from her daughters. From early adolescence on, Beauvoir learned "how to be secretive." She felt this as a clear change in her attitude and summarized the shift in terms of redefining the family's scrutiny as threat:

My relationships with my family had become much less simple than formerly. My sister no longer idolized me unreservedly, my father thought I was ugly and harboured a grievance against me because of it, and my mother was suspicious of the obscure change she sensed in me. If they had been able to read my thoughts, my parents would have condemned me; instead of protecting me as once it did, their gaze held all kinds of dangers for me. They themselves had come down from the empyrean; but I did not take advantage of this by challenging their judgement. On the contrary, I felt doubly insecure; I no longer occupied a privileged place, and my perfection had been impaired; I was uncertain of myself, and vulnerable. All this was to modify my relationships with others.[29]

Although there is much in this that must be regarded as normal in the experience of most developing adolescents, it is important to emphasize that for Beauvoir, as for Sartre, in many aspects of their lives, the leap they made from childhood to adulthood was the leap from the principles of the early nineteenth century to the mid-twentieth century in terms of ideological adjustments and personal values. Both children came from backgrounds where the mainstays of their parents' and guardians' wisdom were, in fact, living a kind of historical afterlife. It is this leap, from, for example in Beauvoir's case, the unreconstructed pieties of the kind of religion represented by her mother and Le Cours Désir to the experimental and situational morality of human relationships that was to characterize her mature views, that gives her work such reserves of exhilaration and freshness. Another case in point is the direct connection between Beauvoir's father's inability to provide dowries for his daughters with the motivation for Simone's family to take seriously her need to work for a living and therefore be educated for a suitable profession.[30]

However, according to her own account, Beauvoir, as an adolescent, was hungry for the education which would prepare her for the work she equally desired. And, indeed, her emphasis on the need for economically viable work for women as the major precondition for and means to their liberation did not waver from the writing of *The Second Sex* to her late interviews with Alice Schwarzer and Hélène V. Wenzel.[31] By her 15th year, Beauvoir had come to an important position regarding both her vocation and the rules and structures by which she would affiliate with others, male and female. These emphasized equality and partnership, ideals which remained consistent in her primary relationships in life. And the friend with whom, and against whom, she worked out these ideas was the first person, outside the family, to make a decided impact on her – her schoolfriend Zaza, the long-dead girl whom the Beauvoir of the late 1950s, writing the first volume of her autobiography, cast as her own doomed *alter ego*, the self that might have been had she not freed herself from the filiative bonds that caught her so tightly in her youth.

In *The Ethics of Ambiguity* (1947) Beauvoir pays particular attention to the importance of the movement from childhood to adolescence in the formation of mature values. The situation of the child, she argues, "is characterized by his finding himself cast into a universe which he has not helped to establish, which has been fashioned without him, and which appears to him as an absolute to which he can only submit."

Under these conditions, all values are "given facts, as inevitable as the sky and the trees," and the child's own inevitable judgment is to consider "values as ready-made things."[32] Adolescence, in contrast, means the discovery of the contradiction inherent in the world of values that had appeared so immutable in childhood. The world is found to be, not solid, but multiply split, and the fissures in the human world represent areas of choice. Thus, "the human character of reality," the fundamental and changeable ambiguity that surrounds the individual, is discovered. It is this, argues Beauvoir, that is "doubtless the deepest reason for the crisis of adolescence; the individual must at last assume his subjectivity." That individual is cast into the ethical world whether they wish it or not; they "will have to choose and decide," will become responsible for the further formation of the self.[33]

Beauvoir's own self-formation, her negotiation through the contradictions of the post-war era in the Paris of the Lost Generation, was played out in relation to three key factors. The first two factors were inextricably bound up with her associations with two contemporaries from her own social milieu; the third factor was the educational process that carried her far away from her family.

The first important relationship, as has been noted, was with Zaza. Beauvoir met Elisabeth Le Coin, the girl she named Zaza in her memoirs, in 1917. Elisabeth was the daughter of a family both richer and more manipulative than Beauvoir's own. She was the third of nine children, and she joined Beauvoir's class in Le Cours Désir after having burned her legs badly in a fire and having been bedridden for a year as a result. Beauvoir was nearly 10 years old when she met Zaza; their friendship continued until Elisabeth's early death in 1929.

In contrast to Beauvoir's own father, who left the Ministry of War to move from a post in his father-in-law's soon bankrupt boot and shoe factory to a series of unsatisfying and usually short-lived jobs as a newspaper advertising salesman, Zaza's father was well placed. An engineer, Maurice Le Coin was director of the Citroën car factory.[34] Zaza's mother, in contrast to Beauvoir's own, was relaxed about her children's behaviour in small things, allowed them astonishing liberties, treated her daughter as a confidante, and ran her household with an assurance that left Beauvoir and her sister both admiring and aghast. The Le Coins were extremely pious and spent part of every summer working for charity in Lourdes. For Beauvoir, the Le Coins were everything her family might have been, and her first reaction to them seems to have been one of collective adoration before they made it clear that they considered the shabby, and later outrageous, Simone de Beauvoir a bad influence on their daughter. This rejection, too, had its uses for Beauvoir.

As for Elisabeth herself, Beauvoir was completely taken with her from the start. An intelligent girl, Elisabeth, like Simone, was "one of the foremost in the class" at Le Cours Désir; a "friendly rivalry" developed between them, and before long, after the many conversations that drew them together during their performance in a school Christmas play, the two girls became known generally as "'the two inseparables'."[35] In the first volume of Beauvoir's autobiography, *Memoirs of a Dutiful Daughter*, Zaza is given a starring role, and Beauvoir's accounts of the importance to her of this friendship startled and surprised both her sister and the surviving members of the Le Coin family. Zaza's function in Beauvoir's highly literary account of her life is, however, clear. She serves as a direct predecessor for Sartre, one who played many of the roles Beauvoir was later to assign to her lifelong partner. The difference, in Beauvoir's telling, is not just that between a man and a woman, though Zaza as an example of what Beauvoir might have been had she not had the strength to resist the dictates of her conventional upbringing is a strong thematic idea in Beauvoir's presentation of her friend. Rather, Zaza's role is that of a soul mate *manqué*, a role that Sartre is later presented as playing with verve, skill, and, most importantly, success.

For Beauvoir, who had never had a close contemporary associate other than her captive sister, Zaza was a revelation, and she elevated her in her imagination to great heights of girlish admiration. Zaza was construed not only as Beauvoir's scholarly equal, but she could also cook, play the piano, perform acrobatic feats, turn out a family newspaper, and, apparently, assume a daring amount of independence with ease. Best of all, Zaza could converse, and she was Beauvoir's first important interlocutor. With Zaza, she could have "real conversations, like the ones Papa had in the evenings with Mama." The girls did not allow themselves any kind of overfamiliarity: they addressed each other formally, as *vous*, and refrained from physical contact.[36] These signs of respect clearly carried a heavy significance for Beauvoir: she and Sartre were also to address each other with the formal *vous* throughout their long lives, an idiosyncratic mark of mutual respect that has been much commented on, and which is clearly related to Beauvoir's experiences with Zaza.

Whether Beauvoir's friendship with Zaza ever reached the stage where the intimate address of the French *tu* would have been appropriate is questionable. But whatever the state of affections between the girls in the 1920s, as opposed to the need for such an association in the patterns of her memoirs, Elisabeth Le Coin served an important imaginative function for Beauvoir, one that allowed her to break out of the enclosure of her family and the deeper enclosure of her self. Zaza was, says Beauvoir, the first person whom she realized she missed when she was absent. Her affection for Elisabeth was, as she herself says, "fanatical," and she admired her friend for her "boyish daring" and her "originality," qualities which Beauvoir cultivated in herself. As she would later do with Sartre, Beauvoir classed Zaza among the "gifted" while she says she saw herself as "merely talented." Reflecting on her own case as an example of the developmental psychology of pubescent girls, Beauvoir noted that she understood that she cast Zaza as a kind of walking miracle, and while this infatuation was real, Beauvoir also understood the amount of self-induced credulity and self-delusion involved in such a schoolgirl passion for a friend. She knew she was something of a willing "victim of a mirage; I felt myself from within and I saw her without."[37] The interplay between Beauvoir's need to cast the first person outside her family she chose to love in a role of glamorous superiority and her understanding that superiority was probably a figment of her own imagination foreshadows the expectations that Beauvoir brought to her later intense emotional relationships. It is also clear that within the structure of her memoirs, what Beauvoir is doing is using Zaza, not only as a foil for Sartre's later entry into the picture, but as a means of attacking the internal deadliness of the class and way of life that she was to reject. Attacking the bourgeoisie was, of course, an old and predictable game. The difference that Beauvoir brings to her rejection of the class of her birth is a gendered one. Her presentation of the gradual snuffing out of the specifically *female* promise that Zaza represented serves as an exemplary instance of what Beauvoir herself might have faced had she not rejected the womanly destiny offered to her by her class.

Whether this literary treatment of Elisabeth fits the reality of the situation scarcely matters. It is safer to read Zaza as an allegorical device than as a faithful portrait of a real young woman. Indeed, taken as a whole, rather than being the soul mate of Beauvoir's adolescence, Elisabeth Le Coin was something of an unknown to Beauvoir. When she was 19 and her need for Zaza's love and approval had, to a large degree, passed, Beauvoir revealed to her friend her former passion for her. Zaza seems to have been amazed by the revelation.[38] Indeed, Hélène de Beauvoir was astonished by the importance accorded to Zaza in her sister's memoirs:

> I had not realized ZaZa meant so much to her, or that my sister continued
> to think of ZaZa for so many years after her death. We knew she was upset
> at the time, but we all assumed it was due to her exaggerated fear of death
> and not specifically of the death of ZaZa.[39]

Elisabeth, indeed, was to die young, in circumstances that will be examined later. What is important to note here is that the closeness, the importance of the association was known to Beauvoir alone.

Beauvoir clearly identified with and admired not only Zaza's intelligence, but her daring and even her instability. Evidence of this instability is stressed by Beauvoir. When her parents interfered with her affection for a cousin when she was 15, Elisabeth thought of suicide. When she was 19, she cut her foot open with an axe to avoid going on a family outing. Beauvoir presents Zaza's extraordinary actions in terms of the classic romantic revolt of a young woman who can do nothing with her intentions to protest against the conditions of her life except to turn on herself in romantic and futile self-destruction. Beauvoir presents Zaza as a young woman driven to the borderline of physical and mental tolerance by the demands of a respectable family who interfered with her wish for a sound education, who forced her to consider marriage as a kind of respectable prostitution, and who stood in the way of her preferred relationships with her friends. She was everything that Beauvoir stubbornly chose not to be and it was crucial for Beauvoir to keep this powerful image of the defeated female in the minds of her readers in the interests both of defending her own hard-won and oppositional choices in life, and of justifying herself to herself. Beauvoir had no intention of presenting her own life as anything but a triumphal progression toward happiness, which overcame formidable obstacles placed in her way. The example of Zaza was her chief representation of the female path that she could have taken, but did not.

If Elisabeth is designated by Beauvoir as the female Other she might have been, the second key relationship of her youth encapsulates another likely womanly destiny that was, most fortunately, avoided. Jacques Champigneulle, Beauvoir's first male romantic object, was her cousin, the son of a man her mother might have married except for her lack of a sufficiently enticing dowry. He represented for Beauvoir the first man onto whom she could project her awakening heterosexual desire (he was, in fact, the only young man the Beauvoir family knew). Jacques is treated in *Memoirs of a Dutiful Daughter* as a temptation and as a threat, and his allure is closely aligned in the structure of Beauvoir's autobiography with the history of Zaza. He was the masculine figure of potential reabsorption back into the milieu that Beauvoir would struggle to escape. Put simply, Jacques was the kind of man she might have married – cultured, artistic in the same limited way as her father, but essentially bourgeois, especially in his attitudes toward women. As with Zaza, Beauvoir's relationship with Jacques seems to have taken place, at least in its greatest intensity, in her imagination. At the age of 17, Beauvoir, in love with love, turned her attention to her largely imaginary courtship with her cousin. Only six months older than she, Jacques's father had died when his son was 2. His mother had remarried, and Jacques and his sister were raised by relatives. At the age of 8, Beauvoir was, she said, dazzled by Jacques's "brilliant compositions, by his knowledge, his assurance."[40] She again stresses her attraction to qualities of brilliance and achievement in her fondness for Jacques. He was a handsome little boy and Beauvoir was delighted to receive his attentions. They regarded themselves as "engaged," and Beauvoir recalled him as the only small child for whom she had any respect.

When she was in her teens, Jacques was the sole young man within reach onto whom Beauvoir could project her need for a suitor. Living alone with his sister and a servant above the family stained-glass business in Montparnasse, Jacques often spent evenings at the Beauvoir apartment. He played the part of elder brother to Simone, treating her as a little girl and helping her with her homework, but also listening to her and taking her opinions seriously.[41]

With Jacques, as with Zaza, as later with Sartre, Beauvoir emphasizes the importance to her of serious conversation, which continued to represent for her the most important bridge between individuals. It was, in fact, chiefly Jacques's intellectual background that qualified him as a suitable screen for her sexual projections. As a day-boy at the Collège Stanislas, Beauvoir's father's college and her adverse point of comparison with Le Cours Désir, Beauvoir credited Jacques with all the glamour of an intellectual mentor and of a desirable (and serious) companion. As she grew older, the points of similarity between them seemed to increase. Jacques's interest in philosophy and his encouragement of her examination successes helped shape her own plans for the future. Searching for a way out of her cage in the mid-1920s, Beauvoir thought she saw in Jacques a potential Virgil to her Dante: "he knew far more than I did about the world, about human affairs, painting, and literature; I should have liked him to give me the benefit of his experience."[42]

Whatever Champigneulle's feelings about his cousin's infatuation with him, Beauvoir's mother was delighted, and relaxed her usual vigilance over her daughter to give her the freedom to make what might have been a most satisfying catch. Françoise was not to know how much of this freedom was intellectual. A large table in Jacques's flat was piled with the latest books and he encouraged Beauvoir to range freely among them. Jacques was, in fact, extremely useful to Beauvoir. He introduced her to the literary and artistic avant-garde of the day. At one of the most experimental and influential moments in the early years of this century, Jacques guided Beauvoir to the work of Cocteau, Mallarmé, Laforge, Proust, and Gide; he took her to avant-garde films and to the director Charles Dullin's latest experimental theatre productions. From Jacques, Beauvoir received her introduction to modern painting: he explained Picasso, Braque and Matisse to her and took her to their exhibitions. Jacques gave Beauvoir her first introduction to the cafés and bars of Paris, and taught her about surrealism. He encouraged her in her own wish to flaunt convention.[43]

Every generation tends to perceive, in one or two novels, the epitome of their own situation. For Zaza, Beauvoir and Jacques, indeed for the intellectual youth of France in general in the 1920s, that novel was Alain-Fournier's *Les Grandes Meaulnes*, which Jacques introduced to Beauvoir, who then tended to use it as a touchstone, reading off the personalities of people important to her in terms of its fictional characters. For her, Jacques was identified with Meaulnes himself, "the perfect incarnation of Disquiet."[44]

The passive, but anarchist, disquiet that so represented the alienated post-war bourgeois youth of France in the 1920s was to infiltrate Beauvoir's own attitudes deeply. With Jacques, as for many, this represented only a phase through which to pass on the way to reintegration with the class of their origins. Beauvoir's analysis of Jacques, in this light, is important for her own development:

> In my view, Jacques was freeing himself from his class because he, too, was suffering from a deep disquiet; what I did not realize was that this deep disquiet was the means which that bourgeois generation was employing in order to effect its own cure . . .

After several years of believing herself in love with Jacques, and believing, too, that marriage with him would almost certainly follow on the strength of their shared interests, Beauvoir discovered, first, his relationship with a former mistress (and this at a time when Beauvoir's own views of sexual behavior still conformed to the convent morality with which she was raised) just at the time she had thought their own relations to be most tender, and, second, to her stupefaction, his contracting of a marriage of convenience to a young woman he scarcely knew but who had the right connections and a large dowry.[45]

Beauvoir presents Jacques's defection, if that is what it was, as a lucky escape for her. She met him just once more in her life after the age of 21: he was 45, prematurely old, a drunk, both physically and financially ruined. He died at the age of 46, of malnutrition, a classic alcoholic's demise.[46] Beauvoir dismisses him from her memoirs curtly: he, like Zaza, is treated as a figure in a social parable, and his death merely the enactment of ultimate justice. With Jacques out of her young life, the temptation of a "proper" marriage would never again arise seriously for Beauvoir. And she saw this as a fortunate development.

Beauvoir clearly thought of both Zaza and Jacques in cautionary terms, yet these two companions, however much she magnified her relationships with them, were of great importance in her own formation. Both were chosen on the grounds of intellectual talent, capacity for mentorship of Beauvoir herself, rebelliousness, and a recognition of her own and their own superiority. All three spoke of writing novels. Beauvoir clearly imagined, at one time, that they were all kindred souls. For all that, the relationships are also characterized by one-sidedness and secrecy. Beauvoir, in fact, knew little of many aspects of Zaza's life, she was not taken into her confidence until her need for an intense adolescent female friendship had passed. She knew even less of Jacques. There is a distinct sense in which Beauvoir was never particularly close to anyone outside her family for her first twenty years, and, like any adolescent, she tended to read other people's actions in naive, schematic, and often contemptuous and self-centred ways. However, for her own purposes, her connections with her bourgeois counterparts in unsuccessful revolt – Zaza and Jacques – were precisely what she needed to show her the way out of the cage. Her own rebellion was already being enacted on different lines, and the major difference between Beauvoir and these failed rebels centered around her education.

Georges de Beauvoir's often repeated lament for his daughters: "'You girls will never marry . . . you have no dowries; you'll have to work for a living'"[47] was, for Beauvoir, a signal for hope rather than despair. When her eleven years of primary school education at Le Cours Désir ended in 1924 when she was 16, her family debated her next step. Her father thought Simone should study law and qualify for a civil service post which would give her security and a pension; her mother thought librarianship would be more genteel.[48] Beauvoir had other ideas. Philosophy attracted her. It had the benefits of being considered with suspicion by her family and with interest by Jacques. Further, Beauvoir had come across an article on a woman who had a doctorate in philosophy, a pioneer, who seemed to have "succeeded in reconciling her intellectual life with the demands of feminine sensibility." Beauvoir was impressed: she, too, wanted to be such a pioneer. Philosophy also suited Beauvoir's own intellectual proclivities:

> The thing that attracted me about philosophy was that it went straight to essentials. I perceived the general significance of things rather than their singularities, and I preferred understanding to seeing; I had always wanted to know *everything*; philosophy would allow me to appease this desire, for it aimed at total reality; philosophy went right to the heart of truth and revealed to me, instead of an illusory whirlwind of facts or empirical laws, an order, a reason, a necessity in everything.[49]

Her ambition was irresistible and in 1925 she began work for the series of examinations for certificates and diplomas which would lead her to her first adult career as a teacher of philosophy in the lycées.

During the four years of her higher education, throughout which Beauvoir progressively moved out of the all-female and Catholic educational milieu of her childhood into the coeducational and humanist atmosphere of the Sorbonne, her family's influence on her steadily waned. At first carefully placing her in institutions where she could not become "infected" with the secularism of the age – the Institut Catholique and the Institut Sainte-Marie de Neuilly – Beauvoir's parents' ability to control her life disappeared almost completely as she gravitated to the Sorbonne and to the Ecole Normale Supérieure toward the end of her studies, where she made important friendships with the outstanding students of her educational generation. Beauvoir's independence, as well as her new friends, led her to experiment with visits to some of the more dangerous and colorful low-life areas of Paris. Yet for all this, Beauvoir's parents' approval of their daughter also grew as her studies progressed. As a woman who had to earn her own living, Simone was, for her father, "a living confirmation of his own failure." If, however, she became "a sort of intellectual prodigy" she would be, not a source of shame, but a "phenomenon who could not be judged by normal standards," someone who "could be explained away as the result of a strange and unaccountable gift."[50] Beauvoir, the dutiful daughter after all, became that prodigy, an adept at passing examinations, one of the youngest students in her cohort.

In 1925, however, this success was still in the future, and the all-female Institut Sainte-Marie was, in fact, a very good place for Beauvoir to begin her studies. Two of the women there were particularly important. The Institut was run by Madame Charles Daniélou, a passionate proponent of education for women, who held more degrees than any other woman at the time in France.[51] Beauvoir's philosophy tutor, Mademoiselle Mercier was an *agrégée* in philosophy, one of the handful of women in the country who had passed the highly competitive examination for this qualification.[52] These were precisely the right intellectual models at the right time for Beauvoir. They set high standards, they demanded respect, their praise had to be earned. The other teacher of importance to Beauvoir at this time was Robert Garric, a young, left-wing, charismatic, and intensely religious lecturer at the coeducational Institut Catholique. Garric taught literature and numbered Jacques among his devotees. Garric, a kind of secular saint, lived in the working-class district of Belleville, to which Beauvoir once made a disappointing sentimental pilgrimage to gaze at his house. Garric represented Beauvoir's first serious brush with the political left. Her father's views, as might be expected from a man who valued Gobineau and Maurras, were that of a far right-wing, paternalist nationalist. Beauvoir's political views seem to have been largely unformed. She was never a particularly engaged political animal, in any conventional sense, but her adulation of Garric directed her toward the anarchist-leftist political stance that she was to retain as her most characteristic political position. Garric ran a series of youth groups, Les Equipes Sociales, dedicated to bringing high culture to the working class. When she was 18, Beauvoir joined one of these groups and spent time in Belleville teaching classes in literature to young, working-class women. It was her first (and largely unsuccessful) attempt at teaching, as well as her first extended view of life outside her own caste. The experience was useful, and the girls of Belleville seem to have taught Beauvoir more than she taught them. Beauvoir identified with her students' need to escape their homes, and used her teaching as an excuse to escape her own. Beauvoir worked hard for both Garric and Mademoiselle Mercier. They encouraged and praised her. The time-honoured process of the gifted student working first to gratify admired teachers before launching into their own work operated well for Beauvoir, whose persona of the dutiful daughter transferred easily to her new role of star (female) student.[53]

The grounding that Beauvoir received in continental philosophy at this time laid the foundations for her thought throughout her lifetime. Kant developed in her a passion for "critical idealism;" Bergson was valued for his theories about "the social ego and the personal ego;" later, Beauvoir read Descartes and Spinoza with attention and admiration, and still later her enthusiasms covered Plato, Schopenhauer, Leibniz, Hamelin, and, especially, Nietzsche.[54] Her final-year thesis, on Leibniz, was written under the direction of the influential neo-Kantian, Léon Brunschvig, and, indeed, Kant remained the most important of Beauvoir's philosophical predecessors.

At the same time as her philosophical studies proceeded with great success and ever more complex development, Beauvoir not only maintained her interest in literature, but felt that her chief ambition – unlike Sartre, who thought of himself as the future author of a grand philosophical system – was not to join the ranks of those skilled in philosophical abstractions, but to write "the novel of the inner life." At the age of 18 she made her first adult attempts at writing fiction: the results were not successful and Beauvoir knew it, but she was nonetheless pleased with the process of writing itself and felt that fiction provided a mode of discourse in which she could put her "own experience into words."[55]

That experience, the experience of a woman moving out of known territory into a singular life for which she had not been prepared and which she had to shape in her own individual way, became more absorbing for Beauvoir as she moved out of her adolescent phase of imaginary friends and into closer contact with other individuals of her own age who liked and valued her as the possessor of a body as well as of a mind and who had no connection with the opinions, favorable or otherwise, of Beauvoir's parents. Beauvoir's education during the period of 1925 to 1929 consisted of more than the formal arrangements bounded by courses, examinations, and solitary literary experiments. In her late teens, and despite her academic success, Beauvoir felt that she was profoundly alone and directionless: "I . . . was breaking away from the class to which I belonged: where was I to go?" Whereas Sartre's life, at this time, was proceeding agreeably and smoothly along familiar gender-based paths for a young man of his class and aspirations, Beauvoir felt, at 18, "condemned to exile," unable to imagine her future, while determinedly abandoning her past.[56] Her lack of perceived direction often drifted her close to despair. However, while scarcely realizing it, and like many intelligent rebels, whether male or female, Beauvoir was in the process of finding her tribe. The search was especially precarious for a woman and Beauvoir scarcely understood the urges behind some of her own actions. Looking for alternatives to the life she knew, Beauvoir, whose reckless streak sometimes outbalanced her caution, sought out adventures. In a spirit of somewhat foolhardy desperation she and her sister, who in some ways was an earlier and even more formidable rebel than Simone, played sexual games in cafés and bars, picking up men and then escaping when matters looked like turning serious. Beauvoir developed a taste for alcohol, and went drinking when she claimed to be teaching in Belleville. The element of risk in these escapades helped her in the process of changing her rather timid ideas about the nature of the relationships between men and women.

In this process of revising her notion of women's possibilities, Beauvoir had a stroke of great good fortune when she was 20 and on a visit to Zaza's family's country estate in the Landes region. There Beauvoir made an important friendship, one that was to serve as a healthy counterbalance to her febrile tie to Elisabeth herself. Beauvoir's new friend was a Polish-Ukranian emigrée, Stépha Awdykovicz, who was working for the Le Coin family as a governess that summer. Stépha was outlandish, exotic, lively, and daring; further, she had a keen sense of her own sexuality, and dared to talk to Beauvoir about sexual matters which Beauvoir's own prudish upbringing had excluded almost from thought, much less mention. Stépha, who had been sent by her parents from Lvov to study in Paris, seems to have simply laughed Beauvoir out of some of her reserve and overpropriety. Certainly, when the young women returned to Paris and kept up the connection that had begun on holiday (indeed Stépha was to be Beauvoir's lifelong friend, and this affection was to extend to Stépha's son, John Gerassi, who became one of Sartre's most perceptive biographers), Beauvoir was both delighted and appalled at the new bohemian set to which Stépha introduced her. Stépha was living with a Spanish painter, Fernando Gerassi, and Beauvoir was shocked and impressed when she discovered that Stépha posed for him in the nude. Indeed, one of Stépha's most important functions for Beauvoir was that she allowed her to take possession of her own body. In contrast to the stiff and untouchable manners that Beauvoir had previously encountered in her ultra-proper milieu, Stépha insisted on touch as a way to signal their friendship. She took Beauvoir's arm in the street, she held her hand in the cinema, she kissed her. Stépha simply accepted the facts of bodily life and refused to be shocked when the two young women caught sight of a pimp being arrested by police in the street. "'But Simone, that's life!'" said Stépha when Beauvoir thought she was going to faint at the sight of the pimp and his whores.[57] Stépha explained men's sexuality to Beauvoir; she talked to her about clothes; she introduced Beauvoir to her bohemian political and artistic friends. She brought, in short, not only daring but pleasure into Beauvoir's life.

Hélène, too, made a contribution to Beauvoir's development. Perhaps even more determined than Beauvoir to make her own way, and, by training as an artist, choosing an even more dangerous path for a woman of her class, Hélène and her friend Gégé Pardo enhanced Beauvoir's network of female friends who were braving the disapproval and difficulties of their novel position to carve out lives for themselves in unorthodox ways. Through these other women, Beauvoir took possession of the world that was, in fact, her adult milieu. The "pacifists, internationalists, and revolutionaries," the artists and experimenters with whom this prim daughter of the bourgeoisie now associated, opened her eyes to new political and aesthetic possibilities.[58] And, as has been noted, the sexual freedom of her new circle was a revelation.

Flirting with an alternative way of living, bowled over by surrealism and Ballet Russe,[59] tempted by bohemianism and revolt, Beauvoir also found her colleagues at the Sorbonne, at which she began her studies in 1928, a source of discovery. In 1928 the pass list in the moral science and psychology examinations was particularly impressive. Simone Weil, who was, with Beauvoir, to be one of the most distinguished philosophers of her generation, headed the list, followed by Beauvoir. After them came the future existentialist philosopher who was to be a close friend of both Sartre and Beauvoir, Merleau-Ponty. Weil was an opportunity missed for Beauvoir. Her revolutionary sympathies were already developed and she clearly thought Beauvoir a bourgeois fool (a judgment with which Beauvoir, many years later, was happy to concur). Their only encounter recorded by Beauvoir concerned a conversation about the relative merits of ontology and revolution. Weil terminated the discussion with the declaration that "'It's easy to see you've never gone hungry'". Beauvoir was annoyed.[60]

Merleau-Ponty was a different matter. A student at the Ecole Normale Supérieure, from a background sympathetic to Beauvoir's own, and still wrestling with the vestiges of his own lost faith, he sought out Beauvoir after the publication of the general philosophy results list. The two students became close friends and philosophical associates. Beauvoir liked him so much that she introduced him to Zaza. Merleau-Ponty, in turn, introduced Beauvoir to Maurice de Gandillac, who took an interest in the state of her faith and who was effusive in his admiration of her brilliance.[61] It was all very flattering for Beauvoir, and her success with these two *normaliens* led her to pursue the acquaintance of another of their number in whom she was especially interested.

René Maheu seemed the most accessible of a threesome which included the future novelist, Paul Nizan, and Jean-Paul Sartre. Their clique, which was composed, in the main, of former students of Alain, the only radical philosopher teaching at the Lycée Henri IV, had a reputation for daring and "brutality," and was the last group of fellow students to remain closed to Beauvoir.[62] Maheu, like Nizan, was married, but Beauvoir was taken with him from her first close contact with him in a talk that he gave in one of Brunschvig's lectures early in 1929. The attraction was physical, and fit in well with her newly discovered talent for pleasure. Beauvoir liked Maheu's face, his eyes, his hair; she found his voice charming. She decided to make his acquaintance and approached him during lunch in the Bibliothèque Nationale. The two hit it off immediately and soon Maheu was writing poems for Beauvoir and bringing her drawings and magazines.[63] Beauvoir found him thrilling, not least for the vigor of his sensuality. Writing her memoirs in 1958, Beauvoir remembered with affection "how proud" Maheu "was of the young red blood pulsing in his veins!" She is careful to note the symmetry of Maheu's liberating influence with that of her new female friend, Stépha. In 1929, Beauvoir said, "I was tired of saintliness and I was overjoyed that [Maheu] should treat me – as only Stépha had done – as a creature of the earth." For Beauvoir, Maheu was "a real man" and "he opened up paths that [she] longed to explore without as yet having the courage to do so." Somewhat to her own surprise, she found herself arguing against the case for premarital female virginity to him.[64] Exactly when the two became lovers is not clear, but what is certain is that Beauvoir remembered Maheu and the first year of her affair with him with pleasure and affection. He helped her complete her own liberation.

Maheu also rechristened her. As Beauvoir, who was known as "le Castor" to her adult intimates, tells the story, it was Maheu who provided her with her new identity. "One day he wrote on my exercise-book, in large capital letters: BEAUVOIR = BEAVER. 'You are a beaver,' he said. 'Beavers like company and they have a constructive bent.'"[65] It was all more apt than either could have known.

For Beauvoir, 1929 was a turning point. In January, she did her teaching practice. She was the first woman to teach philosophy in France in a boys' lycée, at the Lycée Janson-de-Sailly with Merleau-Ponty and Claude Lévi-Strauss.[66] Beauvoir recalled her feelings of exclusion and inadequacy when she used to pass the Collège Stanislas. Now she was in charge of a classroom of just the sort of boys she used to envy. She was an integral member of the next generation of the French intellectual aristocracy in the making. She had discovered pleasure. In one year, racing ahead of her colleagues, she was studying both for her final diploma and for the *agrégation*. Her world had opened up and she felt that she was on the "road to final liberation." She was sure "that there was nothing in the world I couldn't attain now."[67] Her childish sense of superiority had been vindicated as she triumphantly passed her written exams and was taken by her now proud father for a celebratory dinner at the Café Lipp. Beauvoir, indeed a pioneer, was preparing to face the adult world not only with confidence but with an appetite for experience that would never falter.

3

THE OATH

If Beauvoir's romance with Sartre's married friend, Maheu, ripened with the spring of 1929, so too did Sartre's interest in Beauvoir. Maheu was so aware of his friend as a potential rival for Beauvoir's affections that, whenever he was with Sartre, he snubbed her so as not to have to introduce them.[1] Then, at the beginning of May, just after her grandfather died, Sartre saw Beauvoir in a corridor in the Sorbonne dressed all in black, like Simone Jollivet on the day he met her. The sight seems to have moved him, and he made his first gesture in her direction by making a sketch for her of a man surrounded by mermaids, which he signed and labeled "Leibniz bathing with the Monads": the philosopher was Beauvoir's thesis subject, the monads were the seventeenth-century philosopher's elementary units of being. The next day Maheu forgot his caution and presented the drawing to Beauvoir. She may have been charmed by the gesture which flattered both her intellect and her sense of the fantastic, but another month passed and Sartre still had not met her. Yet he was, in his words, "dead set on making her acquaintance."[2] He knew from Maheu what lectures she attended and so could easily have approached her, as was his habit with other women students he wanted to know. With Beauvoir, he was inhibited in a way that had not affected him during his earlier deep infatuation with Simone Jollivet. To understand Sartre's uncharacteristic behaviour, it may help to imagine what Beauvoir may have represented for him.

As a fellow candidate for the philosophy *agrégation*, Beauvoir offered herself for judgment against the same standards as Sartre. And, assuming that she would pass the examination on her first attempt, whereas he had failed, Sartre knew he would never be placed as highly in the competitive academic hierarchy of their generation as Beauvoir. Even if he had genuinely persuaded himself that his examination failure was due to excessive genius, there remained the fact that Beauvoir, age 21, was reaching the academic finishing line nearly three years faster than he. But these were not the only facts regarding Beauvoir that held dangers both for Sartre's traditional male ego and for his perception of himself as, with the exception of Aron, intellectually superior to his peers. Whereas Beauvoir had struggled with poor schools and often contrary parents, Sartre, as he well understood, had enjoyed, from cradle to *agrégation*, nearly every possible educational advantage and encouragement that family, money, connections and masculinity could secure. For months, Maheu brought back reports to Sartre about the amazing Beaver, and so Sartre was well aware of the radical differences in their educational backgrounds and may have wondered what heights Beauvoir might have scaled had she received her fair share of the educational privileges that had been lavished on him and his friends.

In mid-June, after the candidates for the philosophy *agrégation* had sat several days of written exams, Maheu told Beauvoir he was leaving Paris with his wife and would return in ten days to prepare, with Sartre and Nizan, for his oral examination. He said that the men wanted her to join their study group, and that Sartre wanted to take her out. She was delighted with the first invitation and probably would have agreed to the second, having heard Maheu praise Sartre's charm and intellect. But Maheu, fearing the effect Sartre would have on her in his absence and not wanting his own affair with her to end, persuaded Beauvoir to send her sister in her place. On the evening Hélène went out with Sartre, Beauvoir, musing on her future, wrote in her journal: "Curious certainty that this reserve of riches that I feel within me will make its mark, that I shall utter words that will be listened to, that this life of mine will be a well-spring from which others will drink . . . ." Hélène reported back to Simone that "she had done well to stay at home" and that everything Maheu said about Sartre's virtues was "pure invention."[3]

When June ended, Sartre, the womanizer, had still not met the friendly fellow student he was so "dead set" on knowing. There remained a three-week study period before their oral examinations, after which their student days would be over. The legendary meeting between Beauvoir and Sartre finally occurred on the morning of the first Monday in July, when Beauvoir, full of apprehension, reported for the first group study session. This event, surely one of the most important meetings in twentieth-century intellectual history, is invariably related from Beauvoir's point of view, as she was the only one of the couple who wrote about it. But the encounter is perhaps more interesting considered from Sartre's perspective, given that he almost certainly had a much larger psychological investment in the meeting than Beauvoir. When she received the invitation to study with the three *normaliens*, she was pleased because it meant she "would soon be seeing [Maheu] again" and because she "was accepted by his group."[4] Sartre, as an individual, appears not to have figured significantly in her anticipation of that momentous Monday. For Sartre, Beauvoir's arrival in his cluttered dormitory room was the realization of a desire that had lasted throughout the spring and into summer. Presumably, his unreturned interest in Beauvoir had been fuelled, not only by Maheu's reports of the Beaver's critical intelligence and good humor, but also by her attractiveness, which, though not overly apparent in photographs, was much remembered by those who knew her in her youth. Recalling her in 1973, Maheu said "what a heart! She was so authentic, so courageously rebellious, so genuine, and as generous as Sartre. And she was so distinctly attractive, her own genre and her own style, no woman has ever been like her." Nizan's widow, Henriette, who resented Beauvoir, and who still thought, fifty years later, that Beauvoir "should have helped me butter the toast while the two pals [Sartre and Nizan] were together," remembered Beauvoir as

> a very pretty girl [with] ravishing eyes, a pretty little nose. She was extremely pretty, and even that voice, the same voice she has now, rather curious and a little broken, somewhat harsh – that voice added to her attractiveness. I don't think I ever saw her then dressed in anything but black . . . As I said, she was a very serious girl, very intellectual, and these qualities and the black dress actually enhanced her glamour, her unselfconscious beauty.[5]

Beauvoir's confirmed intellectuality and independence of mind, and her rebellious insistence on never covering up these traits or even playing them down when in the presence of men, must have provided a new experience for the young Maheu, whose notion of proper gender roles, Beauvoir said, corresponded to the prejudices of his day.[6] Yet as a well-married man enjoying an affair, Beauvoir posed no fundamental or permanent challenge to Maheu's perceived male role. With Sartre, however, a relationship with Beauvoir which took place on the margins of his existence was, as he surely realized, unlikely. He was romantically adrift, his happy cloistered college days were ending, and his ten-year-old couple relationship with Nizan was disintegrating as his friend settled ever more deeply into marriage and into the Communist Party. Worse still, in November, Sartre faced a year and a half of compulsory military service. More and more his immediate future must have appeared to him as a replay of the most traumatic event of his life: his fall from paradise at 1 rue le Goff into his La Rochelle hell of Others.[7] If he was going to save himself from a second fall, he quickly needed to form a new partnership. His only ready prospect was a woman unlike any he or his time had imagined, and the more Sartre learned of Beauvoir, the more he must have realized that this way out of his predicament would lead him into a relationship as new, original and daring as the woman herself.

Sartre could hardly have been proud of the way he finally engineered his meeting with Beauvoir. That winter, when the sight of Maheu in the National Library had made Beauvoir desire his acquaintance, she had simply followed him into the restaurant, sat down at his table and engaged him in conversation about Hume and Kant. After three months of wanting to meet Beauvoir and having unlimited opportunities to do so, Sartre finally, backed up by Maheu and Nizan, gained his wish as she entered his room, feeling, she says, "a bit scared."[8] Decades later in an interview, she explained why.

> Maheu, Nizan, and Sartre were always inseparable. They came to very few courses because they despised the students at the Sorbonne and the classes there, while the Sorbonne students used to talk about them and say how terrible they were, that they were men without heart, without soul. And of the three, they would say the worst is Sartre, because they considered him a womanizer, a drunk, and a just plain bad person . . . People looked at him with a kind of terror. No one dared say a word to any of them, and they on the other hand refused to lower themselves to talk with anyone else.[9]

Instead of giving Beauvoir a chance to find her feet, the three *normaliens* made her begin by leading them in a discussion of Leibniz, which she did "All day long, petrified with fear . . . ." But each day she went back and soon "began to thaw out." Eventually, dropping Leibniz for Rousseau, Sartre took charge of their revision.[10] Beauvoir recalled that she soon began to see in Sartre

> someone who was generous with everyone, I mean really generous, who spent endless hours elaborating on difficult points of philosophy to help make them clear to others, without ever receiving anything in return. He was also very entertaining, very funny, and forever singing Offenbach and all sorts of other tunes. In other words, he was a totally different person from the one the Sorbonne students saw.[11]

The quartet met every day for two weeks, although they were soon taking afternoons and evenings off. Nizan's wife often joined them and they attended the fun-fair at the Porte d'Orléans and drove around Paris in the Nizans' car. One day Beauvoir slipped off with Maheu and "rented a room in a small hotel in the rue Vanneau" where she was, she says, "ostensibly helping him to translate" Aristotle's *Nicomachean Ethics*.[12]

Maheu was worried that he had failed the written exams. A few days later the results were posted at the Sorbonne, and at the door Beauvoir met Sartre who told her that he, she and Nizan had passed, but Maheu had indeed failed. She knew it meant the departure of Maheu from Paris and possibly the end of their affair.

It was at this time that Sartre famously said to Beauvoir, "From now on, I'm going to take you under my wing," a remark that has been frequently quoted out of context, giving the impression that this paternalistic offer of protection permanently set the terms of their relationship.[13] In fact, however, this appears to be the first time Sartre had spoken to Beauvoir on his own, and the question of forming an intimate relationship with her was not in question. Sartre's remark, reported by Beauvoir in *Memoirs of a Dutiful Daughter*, is placed after accounts of her own bold and manifold confidence. It is presented as a contrast to Sartre's protracted hesitation at making her acquaintance, and has a comic ring, which was undoubtedly intended, as Beauvoir uses Sartre's remark as the introduction to a mocking, sarcastic portrait of his pretensions as a ladies' man:

> "From now on, I'm going to take you under my wing," Sartre told me when he had brought me the news that I had passed. He had a liking for feminine friendships. The first time I had ever seen him, at the Sorbonne, he was wearing a hat and talking animatedly to a great gawk of a woman student who I thought was excessively ugly; he had soon tired of her, and he had taken up with another, rather prettier, but who turned out to be rather a menace, and with whom he had very soon quarrelled.[14]

What Beauvoir later said first attracted her to Sartre was not his pre-emptory claiming of her as a lively disciple, but, far more believably, his qualities of vitality, generosity, warmth, and uniqueness.[15] As she turned from Maheu to Sartre, she turned not to a protector but to an equal.

Sartre understood the choice that Beauvoir now had to make. As he explained to Gerassi, Sartre knew perfectly well that Maheu and Beauvoir's relationship was an intimate one: "Maheu was in love with her . . . And she was in love with Maheu; in fact he was her first lover."[16] But Sartre also felt that Maheu's examination failure had altered the circumstances significantly, and in a way that worked in his favour. With his romantic rival absent, and two weeks of oral examinations about to begin, Sartre pressed his opportunity. With the good impression he had made on Beauvoir over the previous fortnight, he had something upon which to build. He also must have been more determined than ever as he now knew that Maheu's reports regarding Beauvoir's personality and intellectual talent had not been exaggerated. Sartre's plans worked well; he and Beauvoir became recognized companions. Raymond Aron has spoken of the impact Sartre and Beauvoir's partnership had on his own life from the first:

> I think that our relationship changed the day Sartre met Simone de Beauvoir. There was a time when he was pleased to use me as a sounding board for his ideas; then there was that meeting, which resulted in that, suddenly, I no longer interested him as an interlocutor.[17]

In cafés and on the paths of the Luxembourg, where they met every morning, Sartre and Beauvoir began the conversation that would last fifty-one years. They immediately found "a great resemblance" in their attitudes.[18] But more important by far was their discovery that they were committed to similar dreams, whose extravagance required that they think inordinately well of themselves. Their deep-seated ambitions to become writers – in Sartre's case a Great Writer – had sprung directly from their childhood fantasies rather than from adult or even late-teenage assessments of their aptitudes and possibilities. Of course, dreams of literary fame were common in their student milieu, but it was also customary to abandon them when graduation forced their holders out into "the adult world." With Sartre and Beauvoir, however, and especially with Sartre, there is a suspicion that they were psychologically incapable of forgoing belief in their literary futures without suffering mental collapse. If Beauvoir did not aspire to be a philosopher, or believe she was predestined for literary immortality, her aspirations, nevertheless, were more complex than Sartre's and, in total, more immodest. She gradually had become totally committed to leading the life of what today is called a "liberated woman," *and by doing so she intended above all else to win happiness.*[19] Since his days on his grandfather's knee, Sartre could name scores of men who had succeeded in his dream, but Beauvoir could name no woman who had succeeded in hers. The price that women traditionally had to pay for rejecting conventional values was high, and Beauvoir did not intend to pay it. She wanted both liberty *and* happiness, and her relationship with Sartre was to be structured in the light of both of her goals.

For Sartre, taking Beauvoir "under his wing" meant reenacting his most cherished moments with Anne Marie. In Beauvoir's autobiography, she tells how, in their first days together, Sartre led her on searches of the riverside bookstalls for his favourite childhood comics, took her to cowboy films, invited her to share his belief in his destiny as a great writer, and enumerated the adventures he would have when, as he once travelled from A to Z through the *Larousse*, he began his intended world travels – fraternizing with the dock-workers, pimps and white-slavers of Constantinople, the pariahs of India, the monks of Mount Athos and the fishermen of Newfoundland – all the while collecting material for his future masterpieces. To encourage Beauvoir's credulity about his ambitions, Sartre offered her his own credulity regarding her dreams. She had not encountered this before. He also encouraged her to talk about herself and, says Beauvoir, "always tried to see me as part of my own scheme of things, to understand me in the light of my own set of values and attitudes" and explained to her how "I would have to try to preserve what was best in me: my love of personal freedom, my passion for life, my curiosity, my determination to be a writer."[20] But he also warned her, as was his standard tactic with young women he desired, to take care not to infringe his freedom, as, above all else, it was necessary that he remain free to fulfil his destiny as a great writer and great man.[21]

That year's three-man *agrégation* jury agonized over whether to give first place to Beauvoir or Sartre, although all agreed that of the two she was the "true philosopher." One judge held out for Beauvoir, but the others, after initially favoring the woman, decided that since Sartre was a *normalien*, that is, a man, he should receive first place.[22] Meanwhile, Sartre's friendship with the *sorbonnarde* was less than two weeks old and he was already hinting at marriage. Beauvoir, who never forgot his speech to her about preserving his freedom, was too embarrassed to reply, although Sartre was beginning to make her think that perhaps she could both remain true to herself and not have to face the future all on her own.[23] But if they were to have a more permanent relationship it would have to be on some basis other than the conventional one Sartre was suggesting, and, besides, she had yet to be won away from Maheu. She and Sartre were not yet lovers, and at the beginning of August, Beauvoir left him to spend the rest of the summer with her family in the country.[24] There, shortly after arriving, she borrowed money from her cousin, Madeleine, and travelled to nearby Uzerche where she rendezvoused with Maheu and spent three days with him before returning to Limousin.[25] Meanwhile, Sartre was about to redeem himself for the procrastination he had shown when seeking Beauvoir's acquaintance.

Beauvoir was at the kitchen table drinking her breakfast coffee when Madeleine rushed in to whisper that a very short man was waiting for her in the meadows beyond the tower. At once Beauvoir ran out to meet him. Sartre, who was not expected, had arrived the night before and taken a room at the local hotel. Beauvoir described the next four days.

We picked up our discussion at the point where we had left off in Paris; and very soon realized that even though we went on talking till Judgement Day, I would still find the time all too short. It had been early morning when we came out, and there was the luncheon bell already. I used to go home and eat with my family, while Sartre lunched on cheese or gingerbread, deposited by my cousin Madeleine in an abandoned dovecote that stood "by the house down the road"; Madeleine adored anything romantic. Hardly had the afternoon begun before it was over, and darkness falling; Sartre would then go back to his hotel and eat dinner among the commercial travellers. I had told my parents we were working together on a book, a critical study of Marxism. I hoped to butter them up by pandering to their hatred of Communism, but I cannot have been very convincing. Four days after Sartre arrived, I saw them appear at the edge of the meadow where we were sitting.[26]

Sartre, wearing a bright red shirt, "sprang to his feet." Simone's father began a prepared speech in which he asked Sartre to leave the district because his day-long meetings with his daughter were causing a scandal that was damaging the entire family's reputation. When Simone protested, her mother screamed. Sartre spoke, forcefully but calmly, saying that they were in a hurry to complete their philosophical inquiry and so could not postpone their daily consultations, and Monsieur Beauvoir would have to explain that necessity to those whom he thought it concerned.[27] Sartre was victorious: the parents left without replying. Beauvoir realized that "My father and mother no longer controlled my life. I was truly responsible for my self now. I could do as I pleased, there was nothing they could say or do to stop me." And Sartre? It was a performance of which any young man would be proud. He and Beauvoir were now lovers and he proposed marriage. "I told him," said Beauvoir in 1984, "not to be silly and of course I rejected marriage." Sartre, feeling that it was the institution rather than himself that she was rejecting, argued for several days on behalf of marriage, not just in general but as a framework for their immediate lives.[28] Eventually, Beauvoir was forced to remind Sartre of his commitment to maintaining his complete freedom and of his extensive plans for bachelor travels. "I was," recalled Sartre ten years later, "hoist with my own petard. The Beaver accepted that freedom and kept it. It was 1929. I was foolish enough to be upset by it: instead of understanding the extraordinary luck I'd had, I fell into a certain melancholy."[29] When Sartre still remained in Limousin a week after Beauvoir's parents had told him to leave, Beauvoir told him that he had made his point. He should return to Paris and she would join him there. The infatuated Sartre left, clearly uncertain as to whether he had secured Beauvoir's affections on any other than a temporary basis.

Sartre's worries may have been justified. The new lovers were now to be separated for a month and a half. During this time, the pair exchanged letters daily. When Beauvoir returned to Paris in the middle of September, Sartre was absent, and, in the month before his return, there were three major developments in Beauvoir's life. First, she set about clarifying her relationship with Jacques. That her intentions toward Sartre, and perhaps

Maheu too, were still ambiguous is shown by the fact that it was not until after she called on her first love that she wrote of Jacques in her journal: "I shall never marry him. I don't love him any more." A few days later she learned that Jacques was engaged to marry a young woman, not for love but for her connections and large dowry; Beauvoir was shocked by how far she had misjudged him and "heart-broken at the thought of seeing the hero of [her] youth transformed into a calculating bourgeois." But a much greater jolt, as well as a more pointed warning of the dangers that lurked in the conventions of marriage, was the death of Zaza. Earlier that summer, as Beauvoir watched her friends Zaza and Merleau-Ponty fall in love and become engaged, she had thought: "One of my dearest dreams was about to be realized: Zaza's life would be a happy one!" While stiff opposition to the marriage was expected from Zaza's parents, who by custom expected to choose their daughter's husband, Beauvoir thought that at last Zaza believed that she, though a woman, had a right to claim her own happiness. But by the end of summer Zaza's letters to Beauvoir had become despairing: she was caving in to her family's will and she had been forbidden by her mother to see or even write to Merleau-Ponty. When Beauvoir saw Zaza again in late September she "was in a very low state; she had grown thin and pale; she had frequent headaches." Beauvoir "urged her to make a fight for her happiness," and urged Merleau-Ponty, who was worried about his own mother's reactions, to do the same.[30]

Two weeks later Beauvoir received a note from Zaza's mother informing her that her friend was gravely ill, with a high temperature and "frightful pains in the head." From Merleau-Ponty, Beauvoir learned that the day after she urged Zaza to fight for her happiness, she had gone to Merleau-Ponty's flat and in a "confused state" questioned his mother on her opposition to their marriage. Later, Merleau-Ponty arrived, noted Zaza had a high fever and took her home in a taxi. Beauvoir tells how when she next saw Zaza she was in a chapel "laid on a bier surrounded by candles and flowers. She was wearing a long nightdress of rough cloth. Her hair had grown, and now hung stiffly round a yellow face that was so thin. I hardly recognized her." The doctors were uncertain of the cause of her death. "Had it been a contagious disease, or an accident?" asks Beauvoir. "Or had Zaza succumbed to exhaustion and anxiety?" Although decades later she learned more about the events leading up to her friend's death, the tragedy's significance to Beauvoir remains the same: "We had fought together against the revolting fate that had lain ahead of us, and for a long time I believed that I had paid for my own freedom with her death."[31] With those words Beauvoir concluded *Memoirs of a Dutiful Daughter* in which, she later said, "my main desire really was to discharge a debt."[32]

Beauvoir's entire adolescence had been given to her fight "against the revolting fate," and the most important event in her life that autumn was that, when she returned from the country, she began her long-dreamed-of existence as an independent woman. Teaching jobs were assigned nationally and in the first instance were invariably in the provinces, so Beauvoir had decided, *before her first meeting with Sartre*, that she would take her rebellion still further and refuse a post so that she could remain in Paris and have time to write. Ironically, it was Maheu who had pushed her in this radical direction. She says that as she got to know him

> I had the feeling of finding myself: he was the shadow thrown by my future. He was neither a pillar of the Church, nor a book-worm, nor did he spend his time propping up bars; he proved by personal example that one can build for oneself, *outside the accepted categories*, a self-respecting, happy, and responsible existence: exactly the sort of life I wanted for myself.[33]

Moving out of her parents' flat, Beauvoir rented a room in her grandmother's sixth-floor walk-up apartment. Her grandmother treated her "with the same unobtrusive respect she showed her other lodgers." There Beauvoir was free to come, go and entertain as she pleased and, for the first time in her life, had a room of her own.[34] Some private tutoring and part-time teaching of Latin and Greek at a lycée earned her enough money to live.

When Sartre returned to Paris in mid-October, Beauvoir says she had decided that he "corresponded exactly to the dream-companion I had longed for since I was fifteen . . . ." Beauvoir's ideal needs closer examination as its complexity makes it easily misunderstood, and yet it is a factor of crucial importance in determining the nature of the Beauvoir–Sartre relationship. For her dream-companion, who at 15 she still imagined as her husband, Beauvoir says she "had no particular type in mind" but "a very precise idea of what our relationship would be." Rather than be "a man's companion," she would write and have a life of her own. She and her ideal would "be two comrades" who would "be able to discuss everything." This required, she said, that they

> have everything in common; each was to fulfil for the other the role of exact observer which I had formerly attributed to God. That ruled out the possibility of loving anyone *different*; I should not marry unless I met someone more accomplished than myself, yet my equal, my double.

Someone more accomplished than herself, yet her equal, her double: this, of course, is a paradox and Beauvoir tries to resolve it. It emerges that she has two types of equality in mind: equality in terms of achievements and equality in terms of innate potential for achievement, or, as she calls these concepts, equality "from without" and equality "from within." Her ideal companion is her equal from within and must also be nearly her equal from without or else they will not be able to discuss everything. But he cannot, at least in the beginning, be completely her equal from without; it is necessary to her concept of equality from within, which as an idealist she holds as the more important of the two, that he be superior to her from without. At the age of 15, Beauvoir had already understood the basic sociological facts of her female existence. Under the existing state of society and culture, males, *vis-à-vis* females, were favored with better education and, perhaps more importantly, were encouraged at all points to have higher opinions of themselves and to cultivate loftier ambitions. Thus, any man of similar age, class and interests as her own would, as the recipient of decades of this positive discrimination, have had "a flying start" over her. If he failed to show superiority of accomplishments, it would only testify to his inferiority within.[35]

On coming of age, Beauvoir defined herself on the basis of four values which, when situated in the existing sociology of the sexes, came perilously close to being self-contradictory. She was, as explained, committed to the pursuit of happiness, achievement, and independence and, finally, placed great store on having a permanent partnership with her male double. But, without limiting her ambitions, it was extremely improbable that she could find her *alter ego*, and, if she did, then it was also unlikely that he would join with her unless she bowed to social conventions regarding gender roles and unions between the sexes. Alternatively, without an approximation of her ideal relationship, her long-term happiness looked unlikely, and without happiness, her authorial aspirations would also be thrown into doubt. Some may wish to argue that Beauvoir (as she herself once thought) could have steeled herself and got on successfully without her ideal union, but the psychological intensity with which she engaged with Sartre for half a century shows beyond any reasonable doubt that her need for such a relationship was no less integral to her character than her unilateral commitments.

The question remains of how Beauvoir set Sartre up in her mind as her superior from without so that she would not see him as her inferior from within. Even for someone as intelligent as Beauvoir, this could not have been an easy task. If, with his flying start, Sartre had got through the *agrégation* two and half years quicker than Beauvoir, all would have been well, but, of course, the opposite was the case. Sartre had, in fact, at first failed where she had succeeded. Lesser writers would have let this problem with their narrative slip by when composing their memoirs, but not Beauvoir: she attempts to tackle it head on.

> Two years older than myself . . . and having got off to a better start much
> earlier than I had, he had a deeper and wider knowledge of everything.
> But what he himself recognized as a true superiority over me, and one
> which was immediately obvious to myself, was the calm and yet almost
> frenzied passion with which he was preparing for the books he was going
> to write . . . I couldn't imagine living and not writing: but he only lived in
> order to write.[36]

This, as she realizes, scarcely gets Sartre off the ground, but a few pages
later Beauvoir succeeds in lifting him onto a very shaky pedestal. She
focuses on his ambition to be a philosopher – an ambition she did not
share – and claims that he had already brought into existence "a whole
philosophy." For evidence she recalls that the "originality and coherence"
of his philosophical talk "astounded his friends" and quotes from Sartre's
"detailed outline" of his "system of ideas" which had appeared as a letter
in a student journal:

> It is a paradox of the human mind that Man, whose business it is to
> create the necessary conditions, cannot raise himself above a certain level
> of existence, like those fortune-tellers who can tell other people's future,
> but not their own. This is why, as the root of humanity, as at the root of
> nature, I can see only sadness and boredom.[37]

The 24-year-old's letter continues in this pompous, inflated, unsystematic
and undergraduate vein, which, ten years later, a more mature Sartre
himself characterized harshly, if humorously, as his attempt "to translate
craggy, unpolished thoughts into the style of Anatole France." Beauvoir,
however, decided that the young man's verbiage "was positive proof that
he would one day write a philosophical work of the first importance."[38]
Beauvoir's excess of confidence in Sartre needs to be seen as more
than a stereotypical exaggeration of the talents of the beloved. This
kind of extraordinary (indeed unreasonable) leap of faith is usually a
precondition for the creation of such comprehensive works as those
which Sartre intended to produce, and, by believing in Sartre, not only
had Beauvoir found her ideal companion, but Sartre had found someone
whose belief and commitment to his future greatness was as profound
as he had once imagined Anne Marie's to be. But Sartre was getting a
great deal more than a new, improved version of his mother, because,
like Louise, Beauvoir would never believe in Sartre uncritically. Like his
grandmother, Beauvoir was inherently sceptical and questioning. She no
sooner proclaimed the inevitability of his great writer's future than she
began noting his considerable shortcomings, which, if his greatness was
really to emerge, he would first have to overcome. She observed that he
did not write very well, a serious handicap for one who wants to be a

writer, and an even more profound one for the individual who wants to be a great writer. He "refused," she says, "to separate philosophy from literature" and she was "disconcerted by the clumsiness" of his essays. And when she read to him the novel she was writing, he responded in kind: "I was alarmed," says Beauvoir, "to discover that the novel sets countless problems whose existence I had not even suspected."[39]

No part of the Sartre–Beauvoir legend is more central or retold so often as the oath-taking that followed Sartre's October return to Paris in 1929. The story is usually told in something like the following way.

When Sartre returned to Paris, he and Beauvoir resumed their conversations, read each other's manuscripts, played games of man and wife, and began to think seriously about their future together. Beauvoir, being a woman and having waited for Sartre's return like a bride-to-be, surely hoped for marriage. But Sartre, the independent male, demurred and so "it was essentially Sartre who explained to de Beauvoir what the nature of their relationship would be."[40] He informed her that, as he was not monogamous by nature, he could not offer her sexual fidelity or even the opportunity to live with him under the same roof. But, "to temper this revelation" he explained that theirs was an "essential love," whereas any others they had would be only "contingent." The difference was, Sartre said, that they were two of a kind and their relationship would endure for as long as they did. Even so, they signed "a two-year lease," meaning that in that time they would not have contingent love affairs – a promise which, of course, Sartre did not keep. They also agreed that the ideal to which they would peg their essential love was never to lie to or conceal anything from the other, that is, "to tell one another everything."[41] This meant that they would describe in intimate detail to each other their contingent loves, a practice that soon caused the naturally monogamous Beauvoir a great deal of distress.

This story, with its obvious sexual stereotypes, is plausible only if one is both committed to a literal and flat-footed reading of Beauvoir's very literary autobiographies and prepared to forget Sartre's marriage proposals, and Beauvoir's affair with his friend, as well as a wide range of facts regarding the couple's individual pasts and characters.

A review of the timescale and chronology of events will highlight the improbability of the traditional account. Maheu did not leave Paris until some days after the Bastille Day celebrations of July 14, 1929. Beauvoir left Paris at the beginning of August. That gives Sartre and Beauvoir two weeks or less together, in which he, to her astonishment, hints at marriage. In late summer, Sartre appears at Limousin, while Beauvoir has had three days with Maheu in Uzerche. Beauvoir now becomes Sartre's lover and he immediately proposes marriage. She warns him not to be

silly, but for several days he presses for marriage. In the end, she is forced
to point out that his behavior contradicts the claim he made for himself
in Paris, and a few days later, at her request, he leaves. They next see
each other in Paris in mid-October, and Sartre leaves for military service
at the beginning of November. They were left with only two further weeks
together.

The traditional account is rationalized by appeals to sexual stereotypes
and to Beauvoir's memoirs. But her autobiographies are open to other
readings, and, even if they were not, it is difficult to accept the old
story as the true one: too much has become known. More plausible is the
explanation that Sartre, in his self-reported melancholy, came to accept
the fact that Beauvoir was not going to marry him and that the famous
terms they laid down for their relationship were merely the best *he* could
get. Of course, Sartre was not monogamous by nature, but neither was
Beauvoir's father, nor Beauvoir herself, for that matter, and one feels
she would have done almost anything not to undergo a marriage like
her mother's. Besides, Beauvoir, as she says when describing how she
and Sartre came to their understanding, "had broken free of [her] past,
and was now self-sufficient and self-determining," and had established
her "autonomy once and forever, and nothing could now deprive [her]
of it," whereas Sartre had only "more or less shed the irresponsibility
of adolescence."[42] Also, Sartre had two "partnerships" – with his mother
and with Nizan – and with each he had lived in the same room, but there
was no such precedent in Beauvoir's life: she had never lived with Zaza
or with Jacques. Her domestic dream of dreams was to have a room of her
own, and now she had it.

The manner (highly reminiscent of her treatment of Sartre's "take you
under my wing" remark) in which Beauvoir embeds Sartre's request for
"a two-year lease" in a metaphorical incident in *The Prime of Life* should
leave the reader in no doubt as to where Beauvoir thought the balance of
power lay on the occasion of their oath.

> . . . we walked down as far as the Carrousel Gardens, and sat down on
> a stone bench beneath one wing of the Louvre. There was a kind of
> balustrade which served as a back-rest, a little way out from the wall;
> and in the cagelike space behind it a cat miaowing. The poor thing was
> too big to get out; how had it ever got in? Evening was drawing on; a
> woman came up to the bench, a paper in one hand, and produced some
> scraps of meat. These she fed to the cat, stroking it tenderly the while. It
> was at this moment that Sartre said: "Let's sign a two-year lease."[43]

Their relationship, like the balustrade, was to have openings in it; and the
petitioning Sartre, like the miaowing cat caught in the cage, had become

variously swollen through the intermittent attentions of Beauvoir who was now obliging him with the two-year lease, as his army induction, like the evening, was drawing near.

What the usual accounts of Sartre and Beauvoir's initial pledge to each other also tend to devalue is the fact that there were other clauses in their agreement, compared to which their sexual arrangements are of subsidiary importance: for example, they promised they would "tell each other everything" and would never allow anything to "prevail against this alliance." But the putative two-year sexual lease is of interest for what it has for so long covered up. The two years were to be special because in them, says Beauvoir, "There was no question of our actually taking advantage . . . of those 'freedoms' which in theory we had the right to enjoy."[44] This implies that she had agreed to end her affair with Maheu. Indeed, at this point Maheu drops out of her memoirs and their veracity on this matter has always been accepted. But as Beauvoir and Sartre sat on the stone bench in the Carrousel Gardens with the evening drawing on, Maheu was due back in Paris to begin another year studying for the *agrégation*.

A letter, first published in 1990, from Beauvoir to Sartre, which includes a letter from Maheu to her, sheds further light on the events in the Carrousel Gardens in 1929. Dated Tuesday, January 6, 1930, Beauvoir's letter was written only two months after the oath taken in the shadows of the Louvre. The relevant section must be reproduced in full:

I was very annoyed yesterday by a *pneu* I got from the Llama [Maheu's nickname], trying to be wounding in a really infantile way. I shall be very sweet to him on Wednesday, but I find such injustice towards both you and me highly unpleasant. I'm copying out his note word for word, including the significant crossings-out:

"10 o'clock. Forgive me for disturbing you amid all the tender and colourful memories that are doubtless prolonging for you your own dear love's passage. Nevertheless:
Can you be at home on *Wednesday afternoon?* I shall probably arrive at about 3–3.15, since I have a lecture at the Ecole at 1.30. Otherwise (and Sartre must have shown you how unnecessary it was to put yourself out on my account) come and have lunch at Adolphe's *on Thursday at 12.15* (my apologies for not being able, alas!, to take you to Pierre's). I take the liberty of insisting – insofar as I still have any right to do so – that I see you on either Wednesday or Thursday. I have some quite important things to tell you, since it is possible I shall never see you again. For you must understand that I have had my fill of the pretty situation that now exists, as a result of that September of yours and the two months of lying which followed it, and that I deserve something better than the crumbs –

the relations continued out of charity "because I am unhappy" – that you both offer me with such elegance.

Do not be alarmed, at any rate. And, above all, do not write me. That would be the best way not to see me again at all. As things are, I shall tell you quite frankly I am too unhappy to have been able as yet to take any final decision. I shall postpone this, *I promise you* (and my promises I keep), until Wednesday."

I shall assure him, of course, that neither you nor I is prolonging our relations with him out of pity. I want, above all, to try and make him feel my affection for him – and yours too. But I shall tell him, all the same, how astonishing I find this note of his. For absolutely nothing had happened between us from the Saturday when I left him and he was so pleased with me until this Monday – he always said that he accepted this situation, and that what he feared was seeing it change. He, who finds it so easy to reconcile his affections for his wife, for me and for the Humous Lady, is really the last person who can reproach me for loving somebody besides him. I feel, too, that I've put myself out for him more than once, and that these parentheses are pointlessly unpleasant. I was very upset that day at the Napoli and the Café des Sports, when I saw the Llama being so nice after the letter was discovered. I was still a bit upset at the Closeries des Lilas the other day. But this note hasn't upset me at all, because I see it as mere jealousy of a thoroughly disagreeable kind.

How are you, little man? I'm really longing for a letter from you tomorrow. We'll be seeing each other soon, won't we, my love? You promised, so I'm taking good care of myself. I love you. I am, most tenderly, your own Beaver.

S. de Beauvoir[45]

From Beauvoir's letter it can be concluded that she did not end her liaison with Maheu, and the famous two-year lease on sexual fidelity was merely a fiction constructed, not for Beauvoir and Sartre, but for the readers of Beauvoir's memoirs. In fact, the distinction between essential and contingent loves was operational from the first day of the Sartre–Beauvoir pact, and Beauvoir's letter illustrates how the difference, when combined with the full-disclosure principle, actually worked to create their "morganatic marriage." By practicing "translucence" with Sartre and not with Maheu, Beauvoir causes the latter to feel that he is receiving only the "crumbs" left by his rival, despite Sartre's absence from Paris. It is also important to note how far Beauvoir takes her translucence by admitting to Sartre that she was "very upset" when Maheu, after discovering the truth about Beauvoir and Sartre in the letter, did not initially react. Beauvoir's letter also reveals something about her autobiographical technique. *The Prime of Life* begins with Sartre's arrival at Limousin at the end of August 1929, but Maheu is not given a direct appearance in the book until she relates the events of early February 1931, a year and one month after the above letter. Yet, surely, it is this same letter to which Beauvoir now misleadingly refers.

> For some time our relationship had rested on an equivocal basis. He [Maheu] had no intention of admitting what Sartre meant to me, and I had no intention of enlightening him on the subject. Two months previously he had found a letter in my room which made the situation quite clear: he had laughed at the time, but had shown some annoyance too – although he had never concealed the fact that he was highly interested in a girl from Coutances.[46]

If Beauvoir's readers are surprised by her authorial rearranging of the facts, they have only themselves to blame: in the preface to her second volume of autobiography, she clearly states her ground rules.

> At the same time I must warn them [her readers] that I have no intention of telling them everything . . . I have no intention of filling these pages with spiteful gossip about myself and my friends; I lack the instincts of the scandalmonger. There are many things which I firmly intend to leave in obscurity.[47]

Foremost, it seems, among those "many things" to be kept out of the public domain were the precise details of her sexual relations. Contrary to popular trends, sex remained for Beauvoir largely a private and personal matter, and it has become customary to attribute this preference to an innate prudishness fostered by her generation and class. But perhaps rationality was also at work. Perhaps, as someone whose life's work centered on exposing the tyrannies of the Other, she was, in the face of celebrity, committed to preserving her own subjectivity, no less than her privacy. And as someone renowned for her gift for genuine and lasting friendships, perhaps she was also motivated, and not irrationally, to practice discretion out of loyalty to friends. Beauvoir wrote *The Prime of Life* in the late 1950s, and it would have been unhelpful then to Sartre's public image, and to her own, if she had made it known widely that it was her promiscuity, even more than Sartre's, that had determined the terms of their relationship. Furthermore, as shall be seen, the fiction of the two-year lease served to hide more of Beauvoir's sexual activity in the early 1930s than solely her continuing affair with Maheu.

4

BREAKDOWNS AND BEGINNINGS

In November 1929, a month after the New York Stock Market crashed, Sartre's status as a *normalian* cushioned his army induction. He and his classmate, Pierre Guille, were assigned to a meteorology training centre on the outskirts of Paris, where they launched paper airplanes at their colleague and instructor, Raymond Aron. But, in January 1930, this arrangement, which so neatly parodied his student life, ended, and Sartre was sent to a post near Tours without his fellow *normalians*. He was now three hours by train from Paris and he tasted ordinary adult existence devoid of educational privilege for the first time. "Just think," he wrote to Beauvoir:

> the details of my life here are so well regulated, minute by minute, that I know for certain that in eight times twenty-four hours, at 18:15 I will take the same readings, which will begin with the same numbers, after performing the same actions. This will always be mechanical, but I know that the same thoughts will recur, the same hope and despair and all the schizophrenic fabrications which I notice I'm trusting more and more.[1]

If Sartre was tormented by the mechanistic regularity of army life, in Paris, Beauvoir's life radically shifted towards the other extreme of disorder and unpredictability as she settled into a bohemian existence which was as alien to her as Sartre's enforced monotony was to him. For as long as she could remember, Beauvoir's personal reality had been dominated by her preparation for examinations and by the accelerated pace at which she had taken them. She had judged every hour, even those given to recreation, in terms of the series of hurdles society placed between her and the *agrégation*. Now, except for a few hours of simple teaching, her time was free from the demands of any externally defined system of goals. No longer a student, not yet a committed teacher or writer, and actively refusing to honour the rules attendant on her female status, she found herself without a social role to play and cut off from any structured future. In short, Beauvoir had placed herself in a position where she could no longer define herself, even vaguely, by compliance with a preestablished order.

According to the philosophy Sartre later outlined in *Being and Nothingness*, Beauvoir's situation was both psychologically dangerous and epistemologically privileged. Sartrean existentialism begins with the idea that a person is nothing but a series of undertakings which they *choose* in given sets of circumstances; and Sartre's lengthy treatise seeks to explain the consequences of this inalienable freedom. He begins by arguing, in his first chapter, that individuals apprehend their existential freedom with anguish, but, with rare exceptions, evade this apprehension by immersion in routines and values imagined as beyond freedom of choice. However, when an individual is exiled from the everyday world, as Beauvoir was for the two years following her student days, the consequence may be not only anguish, but also fundamental insight into the human condition. It was Beauvoir's experience that Sartre was almost certainly drawing upon when he wrote the following in *Being and Nothingness*:

> . . . there exists concretely alarm clocks, signboards, tax forms, policemen, so many guard rails against anguish. But as soon as the enterprise is held at a distance from me, as soon as I am referred to myself because I must await myself in the future, then I discover myself suddenly as the one who gives its meaning to the alarm clock, the one who by a signboard forbids himself to walk on a flower bed or on the lawn, the one from whom the boss's order borrows its urgency, the one who decides the interest of the book which he is writing, the one finally who makes the values exist in order to determine his action by their demands. I emerge alone and in anguish confronting the unique and original project which constitutes my being; all the barriers, all the guard rails collapse, nihilated by the consciousness of my freedom. I do not have nor can I have recourse to any value against the fact that it is I who sustain values in being. Nothing can ensure me against myself, cut off from the world and from my essence by this nothingness which I *am*. I have to realize the meaning of the world and of my essence; I make my decision concerning them – without justification and without excuse.[2]

Instead of having an imposed schedule of lectures to attend and examinations to take, Beauvoir now faced only the amorphous goal of becoming a published writer. There was no daily network of support, no prescribed way for her to proceed and little likelihood that, whatever way she chose to pursue her ambition, she would be successful. Meanwhile, the pleasures of being young, daring, high-spirited and at liberty in Paris cried out to her, and, within a few months, Beauvoir's literary ambitions were submerged beneath a vivacious hedonism. Moreover, writing for one's self, she found, was not as easy as she had imagined. Ensconced in her own room and faced with blank sheets of paper, Beauvoir discovered that she "had no idea what to write *about*." She says she "found the world's crude immediacy stupefying" and "had no viewpoint of it to present" and, in default, fell into a crude pastiche of her favourite novels: Alain-Fournier's *Le Grand Meaulnes* and Rosamond Lehmann's *Dusty Answer*.[3] From these she patched together a story that was to be her first novel. She abandoned it at the third chapter.

Beauvoir, however, was scarcely short of things to do: that was half the problem. Besides devouring many novels, often in English, while hunched beside the fire in Sylvia Beach's bookshop and lending library, Shakespeare and Company, Beauvoir found herself in demand by an eclectic and shifting assortment of people. With her sister and Gégé she frequented the Jockey Club and La Jungle where, she says, she "made dates and went out with anyone – or almost anyone." Fernando Gerassi, before he and Stépha moved to Madrid to join the struggle to establish the Republic, introduced her into the group of mostly foreign artists, musicians and poets, which included the Delauneys and Tristan Tzara, who drank late into the night at the Café des Arts. Beauvoir attended their gatherings "regularly," and, on at least one occasion, accompanied the group afterwards to the Sphinx, a famous brothel of the day, whose pimps included Henry Miller, and which had the distinction of being the first air-conditioned building in Paris.[4] On other nights Beauvoir entertained various artists and writers from Madrid and Budapest, sent to her by Fernando and by a Hungarian friend. "Night after night," she wrote:

> I showed them around Paris, while they talked to me of other great unknown cities. I also went out occasionally with a young Chez Burma salesgirl . . . whom I found a very likable character . . . We went to dance halls on the rue de Lappe, our faces smothered in powder and lipstick, and we were a great success. My favourite partner was a young butcher's assistant.

"It was seldom," she adds, "that I got to bed before two in the morning."[5]

During Beauvoir's two years "alone" in Paris, Sartre's friends also played a part in her frenetic social life. There was Maheu, of course, who divided his time between the provinces and Paris, but, increasingly, his position as Beauvoir's second lover was threatened by Guille, who, when Sartre was shipped off to Tours, had the good fortune to be assigned to the capital. Since his university days, Guille had enjoyed the favors of a Madame Morel, a wealthy, highly educated, fashionable lady in her 40s, who had been raised in the Argentine and who, with the encouragement of her invalid husband, had taken Guille both as her lover and as the tutor of her two teenage children. Along with a country house and a villa on the Riviera, the Morels maintained a large flat on the Boulevard Raspail, where Guille had his own room. Sartre, though less frequently, had also enjoyed the multifaceted hospitality of Madame Morel. During his absence in the army, Beauvoir became an *habituée* of the Morel household. Several times a week she dined there, or with the household's various members in restaurants, and she often joined Guille and Madame Morel for long drives in the country. In Sartre's absence, Beauvoir quickly found favor with Guille. Madame Morel, far from standing in the way, encouraged the new liaison by loaning the couple her car for trips to Chartres, to the Château de Chaumont and, three or four times, to Tours, presumably to visit Sartre.[6]

On Sundays either Sartre came to Paris or Beauvoir, with an armful of borrowed books, made the long rail journey to Tours, returning by the 5 a.m. train, which arrived at the Gare d'Austerlitz just in time to allow her a few hours of teaching. In the late summer of 1930, Beauvoir took up residence for a month in a small hotel on the banks of the Loire, ten minutes from where Sartre carried out his weather observations using "a sort of miniature Eiffel Tower." He had recently inherited the equivalent of a few thousand dollars from his grandmother, Louise, which, with Beauvoir, he ran through by patronizing expensive restaurants – a lifelong enthusiasm – and by chartering taxis to tour the valley's châteaux. Like Beauvoir, Sartre had plenty of free time, but, in her absence, was not tempted by the local diversions. Instead he applied himself to fulfilling his imagined destiny of becoming a great writer. Yet his vision of the form his greatness would take seems to have remained embarrassingly vague. He initially wrote poetry, which, when read by Beauvoir and Guille, evoked unintended hilarity. He switched to a novel based on the story Beauvoir had given him of Zaza's death. When this was also rejected by his two critics, he moved on to "The Legend of Truth," a "philosophical work" related as a story in the manner of Nietzsche. Beauvoir deplored its "antiquated method" and "the stiffness of his style," but, this time, Sartre persisted with the exercise.[7]

In her autobiography, Beauvoir relates how, during the two years following university, she was more interested in sex than in literary pursuits. In 1929, she says she learned "to take unconstrained pleasure" in her body, and in 1930–31, "separated from Sartre for days or even weeks at a time," she frequently "fell a victim" to sexual desire. Even in the hours immediately after being with Sartre, her appetite for sex could put her at risk. "In the night train from Tours to Paris the touch of an anonymous hand along my leg could arouse feelings – against my conscious will – of quite shattering intensity." The anonymous hand may not have been beyond her powers of resistance, but on other nights of the week she was with men she more or less knew and also liked, and with them, resistance was not so easy, or even attempted. Beauvoir's account of her struggle with her sexuality during this period is full of images which imply that, except on the night train, she was incapable of resisting her "burning pangs of desire" and frequently enjoyed sex with partners other than her mainstays, Sartre, Maheu, and Guille. She speaks of "my tyrannical desires," of "the discrepancy between my physical emotions and my conscious will," of how "my body compromised me completely," and of "lurid desires that . . . struck me with the force of a thunderbolt." She says "I . . . felt my freedom being engulfed by the flesh" and also felt "condemned to a subordinate rather than a commanding role where the private movements of my body were concerned." Despite her unease, this all accorded well with her romantic belief that "The pleasures of love-making should be as unforseen and as irresistible as the surge of the sea or a peach tree breaking into blossom." "My body had its own whims," she writes, "and I was powerless to control them; their violence overrode all my defenses," and although in the morning she would find her desire sated, "by nightfall my obsession would rouse itself once more . . . ." She said she "dared not confess such things" to Sartre, thereby breaking their "policy of absolute frankness." "By driving me to such secrecy," she continues, "my body became a stumbling block rather than a bond of union between us, and I felt a burning resentment against it."[8]

In February 1931, several weeks before Sartre, Guille was discharged from the army. He borrowed Madame Morel's car, and invited Beauvoir to join him on a tour of France. She says she "went straight into a kind of daze, exhilarated . . . at the thought of ten days' tête-à-tête" with Guille. But, two days before their departure, Maheu turned up in Paris without his wife, expecting to spend the next two weeks with Beauvoir. He issued an ultimatum: if she went off with Guille, he would break off their long-standing affair. Beauvoir says she chose Guille, because she "now valued his intimacy more than" Maheu's.[9]

If his partner's sexual promiscuity disturbed Sartre – and, even given the sexual adventurism of their set of friends, it is difficult to think that it would not – neither his nor Beauvoir's writing gives any indication of it. But the fact that they both saw Beauvoir as "well on the road to self-betrayal and self-destruction" was another matter and one about which Sartre's concern appears to have outstripped Beauvoir's. In fact, the manner in which Sartre responded actively to Beauvoir's existential breakdown was thoroughly admirable. Both he and Beauvoir realized that she was in great danger. It was not only that she was losing sight of her goal of becoming a writer, but she was also becoming intellectually passive. In the face of Sartre's mental energies, she, by default, was accepting a secondary status, "that of a merely ancillary being." It was Beauvoir's good fortune – and without it, it seems unlikely that she would be of interest today – that in Sartre she had found perhaps the only male intellectual of his generation in all of France who was not pleased to see his lover lapse into her traditional gender role. Increasingly, Sartre threw his energies into reviving Beauvoir's ambition, her appetite for ideas, and her habit of saying what *she thought* about things. He told her that he especially missed the flood of ideas that used to originate with her which had formed so much of the basis of their discussions. And he warned her not to turn herself into a "female introvert" or a "man's helpmeet." Beauvoir accepted his criticism: "I was furious with myself," she says, "for disappointing him in this way."[10]

Both Beauvoir and Sartre had talked about going abroad separately to work for a few years, but in the spring of 1931 Sartre went straight from the army into a teaching post at a boys' lycée in Le Havre, while Beauvoir accepted a post at a girls' lycée in Marseille which would begin in October. Le Havre was two and a half hours from Paris, but Marseille was 500 miles from the capital and even further from Le Havre. The prospect of this radical separation from both Sartre and Paris fueled Beauvoir's despair and disillusionment. Sartre offered marriage, as, under the rules of the central French educational bureaucracy, this would have entitled them to teaching assignments in the same town. Beauvoir declined. Instead they revised their original pact, agreeing to postpone their separation at least until they reached their 30s. In June their lives brightened when the Gerassis returned to Paris long enough for Stépha to give birth to her son before leaving for Spain. To their satisfaction, the Republic had been established, and Beauvoir and Sartre were urged to come to Madrid to join the celebrations. In August 1931, the French couple, with almost no money but with their generous friends awaiting them, crossed into Spain. It was their first trip outside of France. It set the pattern for their travels together over the next half century, with Beauvoir making all the arrangements and Sartre obligingly trailing after her as she, ravenous for experience, sought to visit every known sight. At the end of September, after six happy weeks, they crossed back into France to take up their new and separate lives as provincial school teachers. At Bayonne, Beauvoir, awash with tears, left the Paris train to pick up the Marseille express.

A week later, from Le Havre, Sartre wrote Beauvoir a long letter describing his new life.[11] He lived in a seedy hotel situated between the train station, the docks and the red light district, patronized the local cafés, taught five hours a day, went to Paris on his Wednesdays off, and walked aimlessly around Le Havre thinking about his new project, an essay on contingency. This was a topic then very much in vogue, rather like "structure" in the 1970s, and on which he and Beauvoir were required to write in their examinations for the *agrégation*.[12] The idea of contingency is a key concept in the phenomenological philosophical tradition to which Sartre and Beauvoir belong. It refers to the proposition that existence of all forms, including physical matter, cannot be identified with necessity. In Sartre's letter to Beauvoir of October 9, 1931, he describes his discovery of a chestnut tree in a local park, which subsequently he would use to famous effect as a symbol for contingency. But the work in which Sartre would utilize this observation from his early days in Le Havre would only appear in the future after a long and sometimes desperate struggle. Meanwhile, he missed his *alter ego*, and told her so:

> My dearest, you cannot know how I think of you every minute of the day, in this world of mine which is so filled with you. Sometimes I miss you and I suffer a little (a very tiny little bit), at other times I'm happy to think that the Beaver exists, buys herself some chestnuts, takes a stroll; the thought of you never leaves me and I carry on little conversations with you in my head.[13]

If Sartre saw himself in Le Havre as exile, Beauvoir's autobiography explains that she treated her arrival in Marseille as a second chance, having, over the previous two years, failed so miserably to establish the self-discipline needed to succeed as a writer. In this Mediterranean city where she knew no one, she hoped to find within her enforced solitude the "greater familiarity with [her]self" that would give her the power to control her daily existence in a manner consistent with her goals. Removed from temptations offered by friends, she set out to free herself, too, from those of her body, subjecting herself to large and regular doses of strenuous physical exercise: tennis in the mornings before teaching, and walking and climbing alone in the rugged hills and mountains above Marseille on her days off. "I never once," she writes, "found myself wondering how to spend my Thursdays and Sundays. I made it a rule to be out of the house by dawn, winter and summer alike, and never to return before nightfall." She continues:

> At first I limited myself to some five or six hours' walking; then I chose routes that would take nine to ten hours; in time I was doing over twenty-five miles in a day. I worked my way systematically through the entire area: I climbed every peak . . . and clambered down every gully; I explored every valley, gorge, and defile.

Beauvoir's solitary rambles developed in her a lifelong taste for "pleasure in driving [her] body to the very limits of endurance," and for the exhilaration of coping with physical dangers, in this case, from the very different threats of rape and of falls.[14] She describes one of her close escapes as follows.

> One afternoon I painfully struggled up a series of steep gorges that should have led me out on to a plateau. The going got steadily more difficult, but I felt incapable of getting down by the route I had clambered up, and so kept on. Finally a sheer wall of rock blocked any further advance, and I had to retrace my steps, from one basin to the next. At last I came to a fault in the rock which I dared not jump across. There was no sound except for the rustle of the snakes slithering among the dry stones. No living soul would ever pass through this defile: suppose I broke a leg or twisted an ankle; what would become of me? I shouted, but got no reply: I went on calling for a quarter of an hour. The silence was appalling. In the end I plucked up my courage and got down safe and sound.

The tough physical regime that Beauvoir initiated to pull herself out of her intellectual torpor worked. Before long, she says, "I had subdued my rebellious body, and . . . I no longer despised myself."[15] Her reading again became purposeful, and, as usual, related to her situation. She was fascinated by the stories, journals and letters of Katherine Mansfield, and by their exploration of the concept of the "solitary woman." Beauvoir also began a new novel, and, unlike the previous one, stuck with it until it was finished. For the first time, Beauvoir felt she did have something to say. There was the previous year's breakdown of her will and self-respect, and her attempt to use Sartre to "release her from the burden of supporting the weight of her own life" upon which to reflect in her writing.[16] And, recalling how she had once tried to use Zaza in the same way, she progressed to the formulation of a general concept which she called "the mirage of the Other." This was the decisive beginning of a lifetime of intellectual originality. The Other became a theme in Beauvoir's new novel and in most of those that followed, but more importantly, in the idea of the Other she, at the age of 23, had found a rich and almost untouched topic for philosophical research. Gradually, over the next two decades, Beauvoir would develop her theory of the Other which would have a profound influence on her career, and an even greater one on Sartre's. The concept, with its supporting theoretical apparatus, would be Beauvoir's gift to the social and intellectual history of the century.

In the autumn of 1932, after her year in Marseille and a summer travelling with Sartre in Spain and Morocco, Beauvoir was posted to a lycée in Rouen, a mere hour from Le Havre and an hour and a half from Paris. It meant that the couple could have Thursdays – their day off – together, as well as the weekends, which they frequently spent in the capital. Beauvoir took a room in a hotel near Rouen's train station and began work on another novel. She was now fully committed to serving a hard literary apprenticeship and was heartened by having Sartre's support near at hand. Beauvoir regarded her new novel, on which she would work for two years and take through several drafts, as purely a learning exercise, intended not for publication but to sharpen and deepen her skills as a novelist.

Sartre was also coming to terms with the fact that, unlike Nizan, he was not going to be an overnight success as a writer. Sartre's friend's account of his year abroad, *Aden-Arabie*, had won critical acclaim. Nizan tried, unsuccessfully, to put his nascent influence to work on Sartre's behalf. He attempted, but failed, to find a publisher for Sartre's student novel, *A Defeat*, and for his philosophical fable, "The Legend of Truth." Sartre kept writing. Gradually, with Beauvoir's help, Sartre's essay on contingency was turning into a novel. Whenever the couple met, Beauvoir read what Sartre had written or rewritten in her absence. He had learned, as she explains, to accept her criticism:

> I knew exactly what he was after, and I could more nearly put myself in a reader's place than he could when it came to judging whether he had hit the mark or not. The result was that he invariably took my advice. I criticized him with minute and meticulous severity, taking him to task for, among other things, overdoing his adjectives and similes.[17]

Their association, however, was not confined to ideas and writing. They shared an avid taste both for gossip and for psychological analysis. Whenever they were together, Beauvoir and Sartre's main topic of conversation was the various people they knew, he especially being fascinated with "reading their characters." Beauvoir found Sartre's "psychological penetration" deeper than hers, but, as she explains in *The Prime of Life*, ultimately her interest in the existence of other people at this time was largely attuned to metaphysical and philosophical speculation, rather than Sartre's more purely psychological interest. In her apprentice novels Beauvoir had undertaken the analysis of "the mirage of the Other," but she says, on a more general level:

> The existence of Otherness remained a danger for me, and one which I could not bring myself to face openly. At the age of eighteen I had fought hard against sorcery that aimed to turn me into a monster, and I was still on the defensive.

Her first method of defence was to pretend that other people "had not eyes with which to observe" her, but, in a few years, Beauvoir would find the confidence to drop her shield so as to explore and analyze the existence of Otherness.[18] The philosophical prize she would gain by so doing would be enormous.

Meanwhile, she and Sartre were also working out their notion of bad faith, which was intended to embrace "all those phenomena which other people attributed to the unconscious mind." Such a concept, which went beyond psychological determinism, was essential to the success of what had become their joint project of developing a philosophy that was both logically coherent – an absolute requirement for the more rigorous Beauvoir – and consistent with their mutual devotion to the idea of human freedom. By observing themselves and the people around them, they set out together to describe self-deceit in all its forms. "We rejoiced," says Beauvoir, "every time we unearthed . . . another type of deception."[19] These discoveries, together with the system of analysis they were slowly constructing, would figure prominently in their future writings.

Sometime during the 1932/3 academic year, Beauvoir and Sartre had an important conversation with Raymond Aron. Aron was in Paris on holiday from Berlin, where, with a year's fellowship at the French Institute, he was finishing a thesis and studying the philosophy of Edmund Husserl. Sartre and Beauvoir were already familiar with the German phenomenologist, thanks to the many-talented Fernando Gerassi, who had studied under Husserl in Berlin, where one of his classmates had been Martin Heidegger.[20] But it was Aron who sparked the couple's interest in Husserl's approach to philosophy and, more importantly at the time, offered Sartre a year's respite from his dull life as a provincial school teacher. It could be arranged, explained Aron to his fellow *normalien*, for them to exchange positions for a year. Sartre was delighted with the prospect. He immediately bought his first book on phenomenology.[21]

With Beauvoir, Sartre visited England at Easter 1933 and toured Mussolini's Italy in the summer, before going on to begin his year in Berlin. Hitler had been Chancellor since January, but there is no evidence that Sartre took special notice of the uniformed thugs who were to be seen in the streets, terrorizing Jews. Sartre, without Beauvoir's hunger for exploration, rarely, it seems, ventured far from the French Institute, where he spent his mornings studying Husserl and his evenings working on a new draft of his novel. He had arrived with high hopes of sexual conquests with the local women, but these quickly evaporated in the face of his inadequate German. He settled instead for an affair with the emotionally disturbed wife of one of the other French fellows. When Beauvoir visited him in February 1934, Sartre introduced her to his lover, thereby forcing Beauvoir to walk the same kind of psychological minefield she had for so long forced him to tread *vis-à-vis* herself, Maheu and Guille. She seems not to have been put off her stride. In Sartre's absence she had rekindled her long-standing affair with Guille. In her autobiography she says that during this period she told Guille everything that happened to her and that "he occupied a most important place" in her life.[22]

When, in October 1934, Sartre returned to his teaching post in Le Havre, he turned over to Beauvoir his German text of Husserl's *Leçons sur la conscience interne du temps*.[23] Having finished her second apprentice novel and with no new writing projects in mind, Beauvoir threw herself into the study of phenomenology. She was an exceptionally fast reader, in contrast to Sartre, who, in an interview in 1972, admitted that he had always been a very slow one. In the same interview he went on to say "it takes me far too long to understand things . . . It takes me a lot longer, for instance, than it does the Beaver here. The Beaver is much faster than I am. I'm more like a snail."[24]

Following his return from Berlin, Sartre soon acknowledged that, as with other philosophical doctrines, Beauvoir's grasp of Husserl's "was quicker and more precise than his own." Every time she and Sartre met, they discussed Husserl's work, and Beauvoir recalls that "the novelty and richness of phenomenology filled me with enthusiasm; I felt I had never come so close to the real truth."[25] Despite its intimidating name, phenomenology's project is the mundane one of the description of experience. Although such an undertaking might seem an obvious starting point to the non-philosopher, it was, in fact, an approach radically neglected by philosophy, not just by French rationalism and German idealism, but also by British empiricism.

Husserl's phenomenological method (called "epoché," which is the Greek word for "bracketing") is to focus on some part of one's experience and then to describe it by "removing" oneself from its immediacy. This psychical distancing is achieved by analyzing away the preconceptions one brings to the objects of one's perceptions. The method's rewards were thought to be twofold: it would reveal the structures of experience, thereby leading to metaphysical insights, and it would bring philosophy back into contact with the real world. But despite these worthy intentions, Husserl's philosophical program very soon reached an impasse. From the Austrian philosopher, Franz Brentano, Husserl adopted the principle of intentionality, which holds that consciousness is always conscious of something, that is, that consciousness intends an object. To this, Husserl added his notion of a "pure" ego or inner self, separated from one's spatio-temporal self, not derived from experience, and residing pristine and soul-like at the "core" of consciousness. The impasse arose when it turned out that it was to this "transcendental ego," and not to the outside world, to which Husserl's bracketing led. The objects of consciousness would be revealed only through the mediation of the transcendental ego and in the form of "essences." Not for the first time, a sincere effort by philosophy to engage with the empirical world had the opposite effect of submerging it in idealism.

It is not clear at what point Beauvoir realized that, ultimately, Husserl's method led her and Sartre away from, rather than toward, their goal of direct philosophical engagement with the world around them. For Sartre, as will be shown in the next chapter, the decisive break with Husserl's method did not come until 1940. Nevertheless, it was in Berlin in 1934 that he began a short essay, "The transcendence of the ego," which, when published in a philosophical journal in 1937, contained a radical revision of the Husserlian position. Sartre had always been strenuously opposed to any notion of a preexistential inner life and so, regardless of what consequences it might hold for phenomenology, was forced to reject Husserl's transcendental ego. "We would like to show here," he writes in his essay, "that the ego is neither formally nor materially *in* consciousness: it is outside, *in the world*."[26] Sartre builds his case for the empirical ego by observing that, when one is fully engaged with the world of objects, such as when reading the words on this page, there is no *I* in one's consciousness. It is only when one stops to reflect on what one was doing that the *I* appears in one's consciousness, and then not as its subject but as its object. Sartre's essay argues that it is by a consciousness reflecting like this on its own activities that one's ego comes into being. These ideas about the structure of consciousness were to prove a critical breakthrough for Sartre and Beauvoir, not so much for the ideas themselves, but because they pointed the young couple in an uncharted direction that would, in time, force them into further original thought.

In literature, Sartre and Beauvoir looked for new techniques that corresponded to the ones they sought in philosophy. Here, in the first instance, they were much more successful. The motto of Husserlian phenomenology was "to the things themselves," and, for their fiction, the couple, especially Beauvoir, had a similar aspiration. In *The Prime of Life*, she recounts how, from the first days of her literary apprenticeship, she was anxious to find techniques that would reduce "the gulf that yawned between literature and life" and "between things and words," that could capture the "here-and-now presence" of reality.[27] Beauvoir explained how, at the beginning of the 1930s, she and Sartre seemed to find a method for reducing that gulf in the fiction of Hemingway.

> Hemingway's technique . . . could be accommodated to our philosophical requirements. The old kind of realism, which described things "just as they were", rested on false assumptions. Proust and Joyce had, each in his own way, opted for a form of subjectivism, which struck us as equally ill-founded. In Hemingway's work the world was still opaque

and externalized, but always examined through the eyes of a particular individual: the author only gave us what could be grasped by the mind of the character he was interpreting. He managed to endow physical objects with extraordinary reality, just because he never separated them from the action in which his heroes were involved: in particular, it was by harping on the enduring quality of *things* that he suggested the passage of time. A great number of the rules which we observed in our own novels were inspired by Hemingway.

Yet Sartre's indebtedness to Hemingway may be rather greater than Beauvoir's remarks reveal. Sartre's *Nausea*, which in 1931 began as an essay and was six years in the making, could not be more different in narrative tone from Hemingway's *A Farewell to Arms* (1929), but beneath the two novels' distinctive surfaces there are some remarkable similarities. There is no hard evidence that either Beauvoir or Sartre ever read *A Farewell to Arms*. But given the avowed importance which the couple attributed to the American's method, and Beauvoir's long-standing frequenting of Shakespeare and Company, the then Paris headquarters of contemporary American writing, it seems unlikely that Beauvoir, at least, had not by 1932 read Hemingway's most celebrated work. That was the year, says Beauvoir in *The Prime of Life*, she persuaded Sartre that he could change his "lengthy, abstract dissertation on contingency" into a novel. She even suggests that she showed him how to plot it, and it is in plot and theme that *A Farewell to Arms* and *Nausea* are so remarkably similar.[28]

Hemingway's tragic novel has as its theme and villain the contingency of existence, which, as in a detective story, is slowly revealed to the outsider hero (a young American caught up in the Italian army) through eternal processes – rain, war, pestilence and childbirth. Lieutenant Henry comes to Italy expecting adventure and believing in the honor and glory of war. Instead he passes time in bordellos before a random shell blows open his knee. Later, as the rain brings his army's advance to a halt in the mud, contingency begins to shed its verbal disguise.

> I had seen nothing sacred, and the things that were glorious had no glory and the sacrifices were like the stockyards at Chicago if nothing was done with the meat except to bury it. There were many words that you could not stand to hear and finally only the names of places had dignity.[29]

In retreat, Lieutenant Henry finds himself about to face a firing squad as he stands beside a river in flood, the novel's ultimate symbol of contingency. He escapes by diving into the water, nearly drowns, but eventually pulls himself onto a bank, where, after his physical struggle with contingency, he experiences nausea.[30] He attempts to forge a separate peace by escaping with his pregnant girlfriend to Switzerland, where, for a while, the rain turns to snow. But when his lover dies in childbirth along with their child, the hero realizes that the villain is not simply war but a more general and inescapable condition.

With a few simple inversions and a failure of nerve at the end, Hemingway's plot is repeated in *Nausea*. Again, contingency doubles as theme and villain, and the narrative proceeds – even more obviously than in *Farewell to Arms* – like a detective story. Roquentin, the outsider hero, is a man past his first youth, who, after lengthy travels, comes to a drab provincial town to do research for a thesis on an eighteenth-century adventurer. Sartre's protagonist gradually sinks, like the Italian army in the mud, into an insular present. As his life sheds its meaning, he loses his belief that he has had great adventures; and, when he decides that there are no adventures, only stories, he deserts his project. Language and routine continue to lose their grip on reality until the full absurdity of the provincial bourgeoisie is revealed to Roquentin, like that of war to Lieutenant Henry. Finally, sitting on a park bench, Roquentin sees a chestnut root, and, by extension, his own body, as manifesting a bare existence beyond all explanation, and, like Lieutenant Henry, he is overcome with a feeling of nausea arising from his desire to escape contingency. His hope for escaping rests with a former girlfriend who is due to visit him, but, when she arrives, her faded self fails to heal Roquentin's psyche. It is then that Sartre evades the force of his own ideas, ending his novel with optimism surging in his hero as he contemplates writing an adventure story that would be "beautiful and hard as steel and make people ashamed of their existence."[31]

Ultimately, however, *Nausea* is driven, not by a Sartrean variation on Hemingway's plot, but by what Beauvoir called its "central purpose – that is, the expression in literary form of metaphysical truths and feelings."[32] Paramount among these was Sartre's phenomenologist's nostalgia to return to things themselves, and, for that reason, it is things, rather than processes, that are the focal points of Roquentin's apprehensions of contingency. *Nausea*'s language and imagery are often overtly philosophical, as befits a novel whose principal aim is to show that a material world exists independently of whatever consciousness makes of it, and to offer, in evidence, Roquentin's nausea as a visceral intuition of that reality.

The text of *Nausea* was already close to its final form when, in 1934, Sartre returned to his school teacher's life, and, under Beauvoir's supervision, began "a scrupulous revision of every single page." Yet now, quite suddenly, the couple's belief in their authorial futures waned nearly to extinction. Beauvoir recalls a day in November when together in

Le Havre they complained "at length about the monotony of [their] future existence."

> Our two lives were bound up together; our friendships were fixed and
> determined to all eternity; our careers were traced out, and our world
> moving forward on its predestined track. We were both still the right side
> of thirty, and yet nothing new would ever happen to us![33]

Individually, Sartre and Beauvoir reacted differently to their falling expectations. Sartre fell into depression and then slid toward madness as his boyhood dreams of the great man's life faded. Beauvoir adjusted, since, as she explained, being a woman then meant that the teaching "career in which Sartre saw his freedom foundering still meant liberation to me."[34] It was her turn to nurse him through a mental breakdown.

Sartre's unstable mental condition did not stop him from writing. In 1935 one of Sartre's professors at the Ecole Normale Supérieure commissioned him to write a book for a series. Sartre's contribution, *L'Imagination* (1936), was a survey of psychological theories of the imagination, made up largely of an elaboration of a thesis he wrote as a student. After quickly completing this book at the end of 1935, Sartre set about trying to create his own theory of the imagination. This led him to an interest in dreams and anomalies of perception, which, in turn, led him to ask to be injected with mescaline at a hospital in February 1935. While under the drug's influence, he reported to Beauvoir by telephone that he was having "a battle with several devil fish." Following his unhappy experiment with the drug, Sartre's depression deepened and he began to suffer from hallucinations, usually of a lobster that trotted along behind him. After an Easter holiday with Beauvoir in the Italian Lakes, not only did Sartre's depression become even worse, but he also often felt himself followed, no longer by a lone lobster, but by a whole army of giant crustaceans. Despite Sartre's clear mental distress, a doctor he consulted refused to provide him with a certificate of leave from his teaching post and, instead, prescribed belladonna which added to Sartre's difficulties. Beauvoir's strategy for dealing with Sartre's breakdown was similar to the one he, a few years previously, had employed against hers. Rather than indulge him with psychological analysis, she "attacked him for the resigned way in which he accepted . . . as a fact" what for him was the intolerable fate of being only a school teacher who at best could only write books that restated other people's theories.[35]

That summer, while Beauvoir climbed mountains alone in France for three weeks, and often slept rough, Sartre went on a Norwegian cruise with his parents. Afterwards, he joined Beauvoir, with whom he walked in the mountains in France for several weeks, all the while followed by his lobsters. But, in the end he declared that he had "sent them packing," and he and Beauvoir returned to their jobs, very much hoping for a better year.[36]

During the previous academic year, Beauvoir had put aside her writing to concentrate on studying philosophy. Now, in autumn 1935, she began a connected series of short stories, while Sartre continued with his psychological studies and his revision of *Nausea*. His turning away from philosophy at this point may have been Beauvoir's doing. In separate interviews given in old age, both Beauvoir and Sartre indicated that she, with familiarity, became so radically disenchanted with Sartre's philosophical talents that she repeatedly advised him to abandon any thought of writing as a philosopher. "I said to him," recalled Beauvoir, "I think you ought to devote yourself to literature rather than philosophy."[37] Sartre remembered her stating the matter rather more bluntly: "Actually, there was a long period during which Simone de Beauvoir advised me against spending too much time on philosophy, saying, 'If you haven't a talent for it, don't waste time on it!'"[38] Presumably, it was not only his time, but her own, that she had in mind. It was clear to her by now that any philosophical projects undertaken by Sartre would call for a great deal of advice and effort from her if they were to reach an acceptable standard of sophistication. She was more certain of Sartre's literary talents and she did what she could to push him in the writerly direction she thought would make his name.

It was also at the end of 1935 that the couple became increasingly entangled with two 19-year-olds, who had entered Sartre and Beauvoir's joint lives in the previous spring as a direct consequence of Sartre's illness. Not wanting him to be left alone with his giant shellfish, Beauvoir recruited two former students – one Sartre's and the other her own – to act as "nurse-companions" for the ailing teacher. Jacques-Laurent Bost, whose older brother was already an established novelist, was befriended by Sartre while still his student at the lycée in Le Havre. Beauvoir describes Bost as "both quick-witted and droll," "with a dazzling smile and a most princely ease of bearing."[39] Bost was also handsome, and, in a few years, would pose a direct and prolonged threat to Beauvoir and Sartre's relationship. Meanwhile, it was the unpredictable daughter of a French mother and exiled Russian nobleman who, at this point, threw Sartre and Beauvoir's relationship into disarray.

Olga Kosakievicz, even more than Bost, "personified youth" for the now 30ish Sartre and Beauvoir. Blonde, impetuous, and categorically opposed to conventional values, Olga had dreamed of becoming a ballet dancer, but, after passing the *baccalauréat* in 1934 under Beauvoir's tutelage, her parents insisted that she begin medical studies in Rouen, a project at which Olga was determined to fail spectacularly. Beauvoir explains that, initially, Sartre valued Olga's company only as a means of deflecting his terrifying lobster hallucinations, but when, in the autumn of 1935, "the crustaceans withdrew they left a kind of vast empty beach behind them, all ready to be filled with new obsessional fancies." Sartre now began to pay "fanatical attention to Olga's every twitch or blink." He was increasingly determined to seduce her, but something rather the inverse of this occurred. In addition to Olga's continuing resistance to Sartre's sexual advances, she gradually seduced him into adopting her system of values, one which Beauvoir says contradicted her own which she previously had shared with Sartre. Beauvoir, in turn, and anxious to agree with Sartre, found herself compromising formerly mutual beliefs. Despite the fact that some of the deepest principles that had held them together were now becoming ever less firm under the influence of Olga, Beauvoir agreed with Sartre that they should annex the young woman to their relationship, and that from now on they would be a trio rather than a couple.[40]

It is usually assumed that Sartre and Olga became lovers, thereby causing Beauvoir much distress, but, as so often seems to be the case when dealing with Sartre and Beauvoir, the legend appears to be the opposite of the truth. In an interview with Sartre near the end of his life, Beauvoir asked:

> Have you sometimes been rebuffed by women? Were there women you would have liked to have certain relations with – women you have not had?
> Sartre: Yes, like everything else.
> de Beauvoir: There was Olga.
> Sartre: Ah, yes.
> de Beauvoir: But that was such a very confused situation![41]

From her letters, published in 1990, it is clear that it was Beauvoir, rather than Sartre, who became Olga's lover, and it is against this background that Beauvoir's account in The Prime of Life of Sartre's jealousy with respect to herself and Olga should be read.[42] Although, under Olga's influence, Sartre slid back toward adolescence, the trio's primary effect on Beauvoir seems to have been intellectual advancement. In discussions, she now not only sometimes found herself opposing Sartre, but, for a period in mid-1936, his jealousy was such that he no longer thought of her as an "ally" and there was a "rift" between them that poisoned the air. Beauvoir says, "I was led to revise certain postulates which hitherto I had thought we were agreed upon, and told myself it was wrong to bracket myself and another person in that equivocal and all-too-handy word 'we'."[43] No less important for Beauvoir was the fact that, at times, Olga treated her in a manner that Sartre had never done. "When she stood apart from me," says Beauvoir, "she looked at me with alien eyes, and I was transformed into an *object* . . . ." Beauvoir and Sartre's relations had always been founded on a reciprocity between themselves as subjects. In contrast, the combination of Beauvoir's current partial alienation from Sartre, and Olga's tendency to "look" at her as if she was a thing, prompted her to think deeply and philosophically about the metaphysical basis of human relations.[44]

In another late conversation reproduced in *Adieux*, Sartre and Beauvoir reveal that, in 1936, despite their differences, they read and discussed the philosophy of Martin Heidegger, but that Sartre's German, even after his year in Germany, was not up to the task, leaving him dependent on Beauvoir's translations.[45] That summer the couple abandoned Olga for travels in Italy, where Sartre's lobster hallucinations recurred briefly in Venice. This seems to have been the final hallucinatory episode of Sartre's breakdown; when the summer was over, he had regained most of his equilibrium.

In the autumn, Beauvoir was assigned to a lycée in Paris, and, to her great pleasure, returned to live in Montparnasse, where she made the Dôme her headquarters. Olga moved into Beauvoir's hotel, and Sartre came to Paris from Laon twice a week. In the capital, the trio resumed its troubled existence, with Sartre still obsessed with seducing Olga. Bost became a philosophy student at the Sorbonne, and Fernando Gerassi, who with Stépha, two years previously, had returned to the Paris art scene and to Sartre and Beauvoir's lives, left to fight the fascists in the Spanish Civil War. Later, Sartre would immortalize Stépha and Fernando in his trilogy, *Roads to Freedom*, but, in December 1936, his plans to become a published novelist were thwarted when *Nausea* was turned down for publication.[46] Beauvoir says that Sartre was "dreadfully taken aback" by its rejection and that, at the end of the year, when they went to Chamonix for skiing, he was still shedding tears over this setback.[47]

Sartre finally admitted defeat with Olga when she and Bost became lovers. Sartre compensated for this further disappointment by trying to seduce every young woman with whom he came in contact. He began a long campaign of seduction on Olga's younger sister, Wanda, who, though not as intelligent as Olga, was perhaps prettier.[48] But, in February 1937, Sartre's attention shifted back to Beauvoir, when, after months of overworking to finish her series of short stories, she fell dangerously ill with a lung infection. In the weeks that followed, her mother and sister, Sartre, Bost, Olga and Madame Morel took turns sitting at her hospital bedside. In the spring, she was well enough to be moved into a hotel, but still too weak to walk across her room. Beauvoir recalled Sartre's care for her at this time with wry tenderness: "The Easter holidays had come; at lunch time Sartre would go and get me a helping of the *plat du jour* from the Coupole, and bring it back to my room, taking short steps so as not to spill anything."[49] She struggled to learn to stand up straight again, and it was a triumph for her when, with Bost and their gay friend, Marco, supporting her by the arms, she managed to walk as far as the Luxembourg Gardens. By April, she was strong enough to travel on doctor's orders to the Midi for convalescence, and while there, at the beginning of May, heard from a jubilant Sartre that Gallimard, after the intervention of Dullin, had accepted *Nausea* for publication.[50] Shortly afterwards, Beauvoir returned to Paris to help Sartre with the novel's final revisions. By mid-July she had her strength back completely, and, after the Bastille Day celebrations, she set off alone for several weeks of mountain-walking in the high Alps around the Col d'Allos. There she had another close encounter with death, when she lost her foothold and fell to the bottom of a ravine. To her astonishment, she had no broken bones, and she picked up her rucksack and continued on her way. Later, in Marseille, she joined up with Sartre and Bost, and the three of them sailed, deck passage, to Greece, where they toured for a month, before making their way back to Paris and to more changes than they could have envisioned.[51]

The year 1938 was momentous for Sartre and Beauvoir. As Europe turned irrevocably toward war, Gallimard's successful publication of *Nausea* in March brought the first secure public recognition that either of them had achieved. It was the first solid evidence that their lives, founded on anticipation of greatness, were not simply enactments of delusions by a pair of eccentric, and ultimately negligible, school teachers. The decade that followed was to "belong," intellectually, to Sartre and Beauvoir to a rare degree. Between 1938 and 1949 they laid down an agenda of philosophical, moral, social and literary concerns central to the culture of the West in the mid- to late-twentieth century. The decade, as they lived it, looked quite different. It was a time, for them, not only of long-awaited success, but also of the shattering of their confident preconceptions, of being forced to realize that their freedom and their subjectivity were decisively bounded by the imperatives of history.

By 1938, Sartre and Beauvoir's lives were settling into the mature patterns which were to provide them with the necessary conditions for their most productive years. Sartre had now returned to Paris to work. Beauvoir was serving her second year at the Lycée Molière in Passy, while Sartre obtained a post at the Lycée Pasteur in Neuilly. The pair had come to terms with their various sexual entanglements (Sartre still had not succeeded with Wanda), and their comradely intellectual liaison, so crucial for their careers as writers and thinkers, as well as mutual daily support, was tested and intact. Sartre's mental illness was under control, while Beauvoir's physical collapse of 1937 was seemingly behind her. The gravity of the European political situation was pushing both of them toward new views regarding the importance of the social responsibilities of the intellectual. The hotel and café life that so suited the pair took on rhythms that were to become, after the cataclysmic interruption of Sartre's wartime internment, characteristic for them. The nucleus of their "family," connected by ties of tutelage, friendship and sexual experience, was expanding. After nearly a decade of ambitious, anxious and, at times, despairing obscurity, Sartre and Beauvoir began to move into their influential place in French culture.

The publication of *Nausea* was a watershed for the fulfilment of Sartre's ambition to become a great writer. His previous failure to come anywhere near success had driven him, literally, mildly insane. In its final form, Sartre's first novel, in the running for the Prix Goncourt and the Prix Interallié,[52] and affectionately dedicated to Castor, attracted instant acclaim. It was followed, less than a year later, in February 1939, by the equally successful publication of *Intimacy* (*Le Mur*), the collection of Sartre's short stories which, together with *Nausea*, established his literary reputation in a remarkably brief period of time.

In contrast, Beauvoir was still floundering in 1938. Her collection of short stories dealing with the lives of a series of French women, *When Things of the Spirit Come First*, had been completed during her convalescence in 1937. Over the previous two years Beauvoir had continued to write, she says, only out of loyalty to her past and because Sartre "pushed" her into it.[53] He thought highly of the result and persuaded her, in the wake of his own success, to allow him to submit her manuscript to Brice Parain, his editor at Gallimard. Beauvoir was so certain of acceptance that she told her parents and friends of her stories' impending publication. Word of her success travelled quickly, especially through the exertions of Beauvoir's proud mother. The humiliation of the stories' subsequent rejection, not only by Gallimard, but also by Grasset, to whom Sartre next took the manuscript, was profound for Beauvoir. She would not allow the collection to appear until 1979. In 1982, Beauvoir described her feelings at the time:

> Two rejections were enough insult, enough humiliation. I was so naive then! If I had only known how many great writers are hurt by repeated rejection of their work, then I might have had the courage to try again with another publisher, but at the time I only believed that my work was inferior, undeserving of public attention. I saw myself as a failure and for a long time viewed myself as unworthy.[54]

Further, Sartre, careful of his own new-found success, warned her against complaining about her treatment by Gallimard. He told her "not to say anything negative about Gallimard, because they were so powerful and he needed them." Beauvoir said that she "kept her mouth shut and swallowed the hurt and told everyone the book was poorly written and because it dealt with silly girls it would probably not have sold anyway."[55] But the rejection was a bitter disappointment to her. While writing the stories, she had decided that they would be the means of her breakthrough into print after years of unpublished apprenticeship writing. She had, in fact, been "sustained by the hope that a publisher would accept them," and buoyed up, too, by Sartre's approval of them. In the midst of Sartre's growing success, the double rejection was a severe blow for her. Worries about the obvious darkening of the international political scene, which she still wished to ignore, but no longer could, combined with her personal frustration, pushed Beauvoir into what she considered "one of the most depressing periods" in her life. She was 30 and she felt her lack of public recognition keenly. Sartre seemed to be drawing uncatchably ahead of her. Word passed among her family and childhood acquaintances that she was a *fruit sec*; her father "remarked irritably that if I had something inside me, why couldn't I hurry up and get it out?"[56]

Beauvoir claimed that it was during a delighted discussion of the success of *Nausea* with Sartre that the idea of both of them becoming "really well-known writers," with "public success, with all its attendant temptations" first entered her mind in a manner that made it seem realizable. In the light of this possibility, she felt that their previous lives looked "rather threadbare." Beauvoir mustered all her forces of stubbornness and refused to admit defeat. Her ambition was clearly fired, rather than extinguished, by her initially depressing rejections. After another climbing holiday in the Alps while Sartre spent time with Anne Marie, a trip to Morocco with Sartre, and a self-centred reaction to the Munich Pact in September 1938, which consisted purely of delight at the avoidance of war, without "the faintest pang of conscience," Beauvoir turned, with renewed commitment, to the composition of her new novel, *She Came to Stay*. Working with utter determination that this new book *would* be publishable, and taking advice while writing it from both Sartre and Parain, Beauvoir saw her fictionalization of the trio with Sartre and Olga as an act which would at last establish her as a writer.[57]

In 1938 and 1939 Sartre and Beauvoir placed their partnership on a new footing which was to remain the basis of their association for nearly the rest of their lives. In the summer of 1939, while they were on holiday in the south of France, Sartre proposed a new understanding to Beauvoir to replace their initial, renewable "contract." As Beauvoir recalled it, this new pact gave her a great deal of pleasure:

> Every October, Sartre and I used to drink a glass of wine to our first pact. But that summer at Madame Morel's villa, we were sitting in the dark one

evening, just the two of us alone, and he turned to me and said, "you know, Castor, we don't need any more temporary agreements. I believe we will be – we must always be – together, because no one could understand us as we do each other." I don't remember what I said except that I was stunned to hear him say this, so out of the blue. Yes, of course, he often wrote such things in letters, but he wrote so many letters each day to so many people that I sometimes thought he made these statements more by rote than by real emotion. I think I just sat there. Then after a while I said, "Yes." I was so happy.[58]

That such a statement of permanent connection was valuable to both Sartre and Beauvoir at this point owed a great deal to the increasing emotional complexity of their association. Their union had never encompassed notions of sexual fidelity; it had, in fact, specifically precluded them. But it is difficult to imagine that Sartre and Beauvoir, pledging their youthful selves to one another, had ever imagined the convolutions of involvement with others that characterized their lives at the end of the 1930s. Yet accounts of the couple's relationship which cast Sartre as a typically sexually unreliable male while Beauvoir patiently tolerated his many affairs are simply misguidedly mapping out conventional notions of male–female relations onto a partnership which was anything but conventional. Since the publication of their letters to each other during this period, it is obvious that Beauvoir not only acquiesced to Sartre's promiscuousness (becoming worried only when Sartre made one of his periodic offers of marriage to another woman, or when another woman threatened to usurp her place as intellectual companion for Sartre), but also joined him in vying for the honours in their mutually enjoyable and voyeuristic accounts of sexual athleticism. Beauvoir's bisexuality, which she repeatedly and publicly had a good deal of fun denying during her lifetime (including during the 1970s when such tastes became fashionable and even politically advantageous in some parts of the feminist movement),[59] was clearly a source of titillation to Sartre. His accounts of his sexual antics, and of the various special tastes of his mistresses, were just as much a source of amusement to Beauvoir. That the two intellectual comrades used other people in deeply suspicious ways is not in doubt. That their promiscuity yielded long-term friendships, and the assumption of equally long-term responsibility in terms of financial, as well as emotional, support for some of the lovers they annexed as members of their surrogate family, is equally indisputable.

Beauvoir's letters to Sartre in the late 1930s indicate a degree of sexual collusion and competition between the pair that shows them both as working out a highly ambiguous desire for joint sexual imperialism which was closely linked with their functions as teachers, and justified in terms of working out a shared life in terms of authenticity which was to remain primary, no matter how many lovers they acquired. That Sartre and Beauvoir's "confessions" robbed their contingent lovers of their sexual privacy, and thus of much of their potential power, was very much to the point. Many of the lovers were treated as semi-disposable, but, when possible, retained as valued friends. Thus, a complex network of former students who had also been, or who were, lovers of one, or both, of the couple, formed the basis of Sartre and Beauvoir's "family." And as the ardor of their own sexual attachment cooled, their intellectual and emotional bonds were tightened by the detailed accounts of their intimate activities that the pair provided for each other.

The nature, and extent, of the pair's sexual colonization of others has only recently, with the publication of Beauvoir's letters to Sartre in 1990, become widely known, and it has prompted distressed responses from many quarters. There is, in general, little difficulty in understanding the rudiments of Sartre's promiscuity. It follows familiar patterns of male desires regarding the formation of harems of attendant women. When one adds to this Sartre's deep-seated attitudes toward his mother, which heightened his fear of being abandoned for a more potent, less ugly, and more adult lover, Sartre's desire to protect himself from female desertion by acquiring a range of women becomes all too understandable. Beauvoir's sexual adventurism and her acceptance of Sartre's in the most open way, as well as Sartre's willingness to share his lovers with Beauvoir, are all less typical (if consistent enough with common generalized variants of modern bohemianism throughout the last two centuries). But the outrage and dismay that has greeted the confirmation of these facts is connected with underlying notions which are still in place regarding the sexual double standard. The furore over the publication of Beauvoir's letters in France is an indicator that such behaviour on the part of a woman is still regarded as beyond the pale.

What remains least conventionally acceptable in all this sexual tangle is the idea that Beauvoir deliberately chose to ally herself with the most significant, amusing, and intelligent male friend and companion she had discovered, rather than looking for something that at least resembled traditional sexual monogamy for women. Sartre and Beauvoir chose writing, mind and friendship as the most important indigents of their association, and they chose these factors over promises of sexual fidelity for which they substituted a code of honesty in and reportage of sexual relations with others. Their ideal was, as Beauvoir put it in a letter in 1940 to Sartre, "our old (old but still true) idea of ethics without deserts – of grace, and the gift," an ethics reminiscent of Simone Weil's, one which the couple considered valid and matched with precision to the immediate situation of the ethical actor.[60]

However they justified their behavior to themselves, this era of shared lovers was, in its way, remarkably productive for Sartre and Beauvoir as a couple moving out of the phase of their first infatuation but eschewing marriage. It worked in powerful and at times, perverse ways, as will be shown in the next chapter, to generate material for the production of some of their major writing, besides drawing together members of their surrogate family. It fed their mutual delight in gossip, intimate scandal, and emotional thrills. It allowed them to conceive of themselves as daring bohemian rebels rather than as sinking into the torpor of bourgeois respectability to which they feared their profession as teachers relegated them. It provided them with an entourage of intimately annexed inferiors and students who could witness and validate their own singular bond. And it identified them, further, to themselves, as two of a very special kind. They had, as Beauvoir put it, "the identical sign on both our brows,"[61] and the rules they established for their association were ones they neither recommended to others nor allowed others to judge during their lifetimes outside the closed school of the family they thus constructed.

On September 1, 1939, Germany invaded Poland; the United Kingdom and France declared war on September 3; Sartre was called up on September 2. Beauvoir and Sartre were to be separated from the time that Sartre left Paris to join the meteorological unit of an artillery division in Alsace, near Strasbourg, on September 2, 1939, until his reappearance in Paris after his incarceration as a prisoner of war in March 1941. They saw each other only three times during this period: twice during Sartre's leaves in February and April 1940, and once during a clandestine visit by Beauvoir to Sartre in Brumath in Alsace in November 1939.

Beauvoir reported that, in the early summer of 1939, she was still unwilling to face the reality of the coming war, though Sartre tried to prepare her for its inevitability. In late July, Beauvoir met Sartre and Bost, both full of foreboding, in Marseille. A few days later they happened to meet Nizan, who had left the Communist Party and was sailing for Corsica with his family. Nizan, who was killed in the war and vilified by his former fellow communists, was never seen again by them. Alarms regarding mobilization sent Sartre and Beauvoir back, after a stay at Juan-les-Pins with Madame Morel, to an eerily empty Paris, reached with difficulty via already disrupted trains. Wandering around the city, tense with anxiety about the future, escaping only temporarily into Sartre's childhood fantasy world of Hollywood westerns in the evenings, they comforted themselves with the shared conviction that the war would, at least, be short, and that it would mean the final defeat of fascism by the democratic countries of Europe. This would bring about, they hoped, a general upsurge in socialism in Europe. Construed this way, the war became imaginatively bearable.[62] And, although Beauvoir was frightened for Sartre's safety as a soldier, her major worry was about the misery of their probable separation. Sartre tried to convince her, and himself, that, as a meteorologist, he would be in little danger. On the train that took him to his first military destination at Nancy, he resolved to fraternize with his fellow soldiers, finished reading Kafka's *The Trial*, and began *In the Penal Colony* as preparation for the fully lived surrealism of the war that was about to engulf him.[63]

5

THE TRUE PHILOSOPHER AND THE MAN WITH THE BLACK GLOVES

Besides maintaining that Sartre was her first lover, that she was monogamous and exclusively heterosexual by nature, Beauvoir insisted, to the end of her days, that she had no influence on Sartre's philosophy. Initially, her disclaimer merely confirmed the almost universally held belief that originating important ideas is an exclusively male prerogative, and Beauvoir was portrayed – and often still is – as serving as a mere intellectual midwife and wet nurse for the amazingly fecund Sartre. According to the legend, *all* the philosophical ideas found in Beauvoir's work originated with Sartre. Supposedly, by the late 1930s, the couple's initially symmetrical intellectual relationship had been transformed, without strife, into one of perfect asymmetry. But of all the parts of the Sartre–Beauvoir legend, this one was giving the most problems by the end of their lives. By the 1980s faith in the primacy of men's intellects no longer looked secure as a cultural certainty. A growing number of individuals, mostly women, engaged in a thoroughgoing critique of this assumption of universal masculine ownership of ideas.

In this changed cultural ambience, old assumptions about the Beauvoir–Sartre legend were questioned. It had long been known that Sartre wrote most of *Being and Nothingness* upstairs in the Café Flore with Beauvoir sitting next to him. In the 1980s, questioning minds began to wonder aloud whether, perhaps, Beauvoir's role in those joint sessions had not been the traditionally feminine ones of secretary, editor and moral support. But the most serious threat to the legend arose in connection with the idea of the Social Other, a concept credited to Sartre and as important to the second half of the twentieth century (through its use by liberation movements and in the analysis of social oppression) as the idea of the Unconscious was to the first. The concept of the Social Other is central to Sartre's later work, especially to his *Critique of Dialectical Reason*, and first appeared in his writing in *Saint Genet*, written between 1950 and 1952. But the concept was already to be found fully developed in 1949 in Beauvoir's *The Second Sex*, where Beauvoir used the concept of the Social Other as the mechanism that explains the social oppression of women. Furthermore, in her still earlier book, *The Ethics of Ambiguity* (1947), she can be observed seriously developing the concept.[1] Once the possibility of important ideas originating with women was no longer ruled out categorically, then the set of facts mentioned above decisively challenged the basic outlines of the Sartre–Beauvoir legend. Interviewers began to press Beauvoir on the matter, urging her to admit that she had not only influenced "Sartre's philosophy," but had done so in a major way. Despite all the public facts to the contrary, Beauvoir stuck to her version of the legend: "I had no philosophical influence on Sartre."[2] She continued to deny both her influence on Sartre and her own philosophical originality even after Sartre died, and she was still denying it at the time of her death in 1986.

This is a matter that has greatly perplexed and disappointed Beauvoir's admirers. It also has remained a profound mystery. Why would Simone de Beauvoir, of all people, want a woman's central contribution to the stock of philosophical ideas credited to a man? It is now possible to answer that question. Two sets of documents, which Beauvoir left to be discovered after her death, reveal facts which provide a solution to the mystery. The Social Other, it seems, is merely the tip of an iceberg of intellectual indebtedness of Sartre to Beauvoir. This chapter will attempt to measure the dimensions of this colossal debt. Doing so requires close reading of key passages from *She Came to Stay* in order to chart the development of Beauvoir's philosophical views. Some of the ideas necessary to analyze this development are intrinsically difficult. But the prize is the correction of one of the great legends of this century, one which also brings into focus what is bound to be a central social issue in the century to come.

" 'You both have so many ideas in common,' said Xavière, 'I'm never sure which of you is speaking or to whom to reply.' "[3]

Xavière's observation is addressed to Françoise and Pierre, the other two members of the *ménage à trois* in Beauvoir's first published novel, *She Came to Stay* (1943). Because Beauvoir based Françoise on herself, Pierre on Sartre, and Xavière on their friend, Olga, critics customarily read the novel as an approximation (Françoise murders Xavière) of the real-life trio's relations, and feelings for one another. But *She Came to Stay* is also packed with what, in a male-authored novel, would be called philosophical ideas. Consider the following scene where the trio, still in its early days, is in a nightclub where Françoise calls their attention to a young woman sitting at another table.

> She was staring, as if hypnotized, at her companion. "I've never been able to follow the rules of flirting," she was saying. "I can't bear being touched; it's morbid."
>
> In another corner, a young woman with green and blue feathers in her hair was looking uncertainly at a man's huge hand that had just pounced on hers.
>
> "This is a great meeting-place for young couples," said Pierre.
>
> Once more a long silence ensued. Xavière had raised her arm to her lips and was gently blowing the fine down on her skin. Françoise felt she ought to think of something to say, but everything sounded false even as she was putting it into words.

And a few minutes later:

> The woman with the green and blue feathers was saying in a flat voice: ". . . I only rushed through it, but for a small town it's very picturesque." She had decided to leave her bare arm on the table and as it lay there, forgotten, ignored, the man's hand was stroking a piece of flesh that no longer belonged to anyone.
>
> "It's extraordinary, the impression it makes on you to touch your eyelashes," said Xavière. "You touch yourself without touching yourself. It's as if you touched yourself from some way away."[4]

These parallel passages exhibit and contrast four ways of experiencing the human body, distinctions which were later partially echoed in Ryle's *Concept of Mind* (1949), and which provided a highly original way around the classic mind/body problem in Sartre's *Being and Nothingness*. The fourfold distinction is as follows: there is my body as part of my lived subjectivity, that is, the instrument by which I am in-the-world; there is my body as seen by others; there are the bodies of others; and there are bodies construed as purely physical objects (the "body" of Cartesian dualism). Beauvoir's nightclub scene self-consciously shifts back and forth between these four philosophical points of view. The third-person narrator presents the bodies of others: the young woman staring and speaking to her companion, the woman with feathers looking at a man's hand, Xavière doing things to her arm, the feathered woman conversing, and Xavière touching and talking to herself. But Françoise's consciousness is also presented, and hence her unself-conscious experiencing of her body as her means of thinking, hearing and speaking. These two modes of experiencing the human body (as an object belonging to another subjectivity and as part of one's own subjectivity) are found in most narratives; it is Beauvoir's structured weaving of the other two modes through the scene that shows her philosophical intent. The two women coping with male flirtations are contrasted by the way they respond to the possibility of experiencing their bodies as objects of another's subjectivity. The woman with the emblematic feathers in her hair decides to experience her arm as a mere thing impersonally related to her consciousness. Similarly, the reader's attention is drawn to "the fine down" on Xavière's skin. Beauvoir's description of touching one's own eyelashes illustrates the unbridgeable difference between experiencing one's body as the instrument of one's subjectivity and experiencing it as an object. In *Being and Nothingness*, Sartre also uses the example of touching oneself to introduce his discussion of the body and its modes of being.[5]

As well as profoundly extending ideas about the possible relationships between minds and bodies, Beauvoir's nightclub scene also illustrates the concept of bad faith, another philosophical idea fundamental to the system of thought later posited by Sartre in *Being and Nothingness*. Implicitly, the woman whose hand has been pounced upon and who has decided to pretend that her arm is an inert piece of flesh is in a quandary. Although she does not welcome the man's desire, she also, perhaps, does not wish to shatter her impression that he "desires" her conversation. Her response is to dissociate the two sides of her human reality, which is that she is simultaneously an object and a subject. According to Beauvoir and Sartre, bad faith occurs when individuals refuse to coordinate these two dimensions of their existence. In terms of this study, Beauvoir's example of bad faith is of particular interest. No philosopher since Plato has been more successful than Sartre at providing vivid illustrations of concepts, and one of Sartre's most famous successes is his unattributed use in *Being and Nothingness* of Beauvoir's illustration of bad faith outlined above.

In the traditional reading of the Beauvoir–Sartre relationship, coincidences of ideas and imagery between their works are taken as evidence of female intellectual dependence on masculine thought. There is, however, in the present case, a problem with the orthodox interpretation, and, by any standard, it is a formidable one: Beauvoir wrote *She Came to Stay before* Sartre wrote *Being and Nothingness*. Although both works were published in 1943, it only became known in 1990 that, five months before beginning *Being and Nothingness*, Sartre read what was over half, and very close to the final version, of *She Came to Stay*. Because, as will be explained in the second half of this chapter, unreliable information was provided about how much of Beauvoir's novel existed for Sartre to read on his army leave in February 1940, some of the recently revealed facts must be stated. They come from two sources: Beauvoir's *Journal de guerre* (1990) and her *Letters à Sartre* (1990). Beauvoir's writing of *She Came to Stay* can be traced in both documents, but the letters are especially revealing. Between October 5, 1939, and January 22, 1940, Beauvoir commented on her progress with the novel in over thirty letters to Sartre. They show that he had previously read and discussed a draft (200 manuscript pages) of what was to be approximately the first 40 per cent of her novel. On December 7, Beauvoir, having drafted another 300 pages, wrote to Sartre: "Since yesterday, I've been revising the novel from the beginning. I've had enough of inventing drafts; everything's in place now and I want to write some definitive stuff. I'm enjoying it enormously, and it seems terribly – quite seductively – easy."[6]

Her letters describe, chapter by chapter, her progress with her "final version" through December and January. By December 29, she had 60 pages in "final draft," by January 3, 80 pages, and by January 12, 160 pages. The revision proceeded faster than she expected, twice causing her to revise upward her promise to Sartre of how much of the final draft she could show him in February. On January 17, she wrote: "I really think you'll heap me with praises when you read my 250 pages (for there'll be at least 250 . . .)".[7] Beauvoir's war journal says that Sartre arrived in Paris on February 4, spent the next morning reading *She Came to Stay*, and had at least *seven* more sessions of reading her novel before he left on February 15.[8] The key question in noting these dates so closely relates to the material in the first half of Beauvoir's novel that must have been of special, indeed extraordinary, interest to the future father of French existentialism.

That *She Came to Stay* has not been generally recognized as a philosophical text seems due to more than just the fact it was written by a woman.[9] First, unlike Sartre's fiction, its philosophical content is so deeply integrated into its narrative structure and handled with such finesse that its very existence is easily overlooked. Second, the novel invites and has received three other major readings: initially as a sociological study of up-market bohemian Paris; later, after Sartre and Beauvoir became famous, and it became known that the novel's major characters were based on them and their friends, as an account of their relationship; and today, with the rise of feminism, as a comparative study of three non-traditional women coping with a male-dominated world. These kinds of readings are all, in their ways, illuminating. However, once the novel's structural base is understood, philosophical ideas begin to leap from nearly every page. The work, in fact, articulates a philosophical system that in its basic structure differs almost not at all from the one found in *Being and Nothingness*. Through skilful orchestration of Socratic dialogues, imagery, dramatizations and third-person narration focused on characters' conscious-nesses, Beauvoir had already produced a full statement of "Sartrean" existentialism by 1940.

In the opening eight pages of *She Came to Stay*, Beauvoir, in addition to beginning the story and fleshing out two of its major characters, blueprints most of the metaphysical architecture on which Sartre was later to build *Being and Nothingness*. Even if this philosophical edifice had been Sartre's creation, Beauvoir's presentation would deserve attention because it is so elegant, convincing, and economical that an easier way into the difficult ideas found in Sartre's longer and more famous statement of the philosophy can scarcely be imagined.

The novel opens with a description of Françoise's consciousness of her surroundings as she – and this is symbolically significant – works in a deserted theatre with the Bost-like Gerbert, revising a script of *Julius Caesar*. "The typewriter was clicking, the lamp threw a rosy glow over the papers . . . 'And I am here, my heart is beating.'"[10]

Beauvoir's seven-word definition ("I am here, my heart is beating") of what Sartre was to call being-for-itself is a densely packed revision of classical and modern philosophical positions. Six points must be noted. By beginning with consciousness, Beauvoir is founding her novel's philosophy on a basis different from that of Heidegger's "existentialism," which begins with being. Second, the consciousness is intentional, that is, it is of something other than itself – for example, the typewriter clicking, the rosy glow – and so, by its very nature, is connected to the world. Third, the kind of consciousness represented is prereflective and therefore breaks with the Cartesian tradition, which begins with consciousness reflecting on itself. Fourth, the word "here" places the conscious individual in the physical world. Fifth, "I am here" indicates that the consciousness is aware of being conscious. Finally, the addition of "my heart is beating" identifies the conscious being as a psychosomatic unity, and, in so doing, opens a new chapter for philosophy.

Having begun her ontology with individual consciousness, Beauvoir faces the problem inherent in this approach: demonstrating that an external world really exists. To succeed where others have failed, she must show both that Françoise's consciousness is not reducible to its perceptions, and that the external world is not reducible to her conscious-ness. Beauvoir's first move is to show that Françoise has certain powers over the content of her consciousness:

The ashtray was filled with stub-ends of Virginian cigarettes: two glasses and an empty bottle stood on a small table. Françoise looked at the walls of her little office: the rosy atmosphere was radiant with human warmth and light. Outside was the theatre, deprived of all human life and in darkness,

with its deserted corridors circling a great hollow shell. Françoise put down
her fountain pen.

"Wouldn't you like another drink?" she asked.

"I wouldn't say no," said Gerbert.

"I'll go and get another bottle from Pierre's dressing-room."[11]

The theatre is currently outside the range of Françoise's perceptions and
yet the narration implies that various images of it have entered her
consciousness. It is also clear that Françoise has imagined the possibility
of her office as differing in one respect from how she perceives it, that
is, as containing a non-empty bottle of whisky. Thus, with respect to
the contents of her consciousness, Françoise's imagination gives her the
power to negate her immediate perceptions. It is by such a negation
that she experiences whisky as a *lack* and that the remembered bottle in
Pierre's dressing-room suddenly acquires meaning as a possibility. Gerbert
concurs and an action – a surpassing what-is toward what-is-not – which
will further change Françoise's perceptions, is freely decided upon. With
the exception of time, Beauvoir has now marked out the seven aspects
of consciousness – intentionality, self-awareness, psychosomatic unity,
negation, lack and possibility – which she explores at length in her novel
and which Sartre later reelaborated in *Being and Nothingness*. In addition to
the empty bottle and the unfinished manuscript, Beauvoir also lists a series
of "negative experiences" (later named *négatités* by Sartre), each implying
a transcendence of pure perception: "stub-ends," "outside," "deprived,"
"in darkness," "deserted and hollow shell," the last a recurring symbol
of nothingness in her novel. Thus, in this brief passage, Beauvoir has
shown that Françoise's consciousness is not merely a passive receptacle
for perceptions of a material world, and is, therefore, an ontologically
primitive mode of being. Furthermore, if, when Françoise leaves her
office, she perceives and finds the putative theatre and whisky bottle,
then Beauvoir will have gone some way toward demonstrating that a
world of material things really exists. In this way Beauvoir hopes to find
the elusive middle ground between materialism and idealism.

And she does in her next paragraph.

She [Françoise] went out of the office. It was not that she had any
particular desire for whisky; it was the dark corridors which were the
attraction. When she was not there, the smell of dust, the half-light, and
their forlorn solitude did not exist for anyone; they did not exist at all. And
now she was there. The red of the carpet gleamed through the darkness like
a timid nightlight. She exercised that power: her presence snatched things
from their unconsciousness; she gave them their colour, their smell. She
went down one floor and pushed open the door into the auditorium. It
was as if she had been entrusted with a mission: she had to bring to life

this forsaken theatre now in semi-darkness. The safety-curtain was down: the walls smelt of fresh paint: the red plush seats were aligned in their rows, motionless but expectant. A moment ago they had been aware of nothing, but now she was there and their arms were out-stretched. They were watching the stage hidden behind the safety-curtain: they were calling for Pierre, for the footlights and for an enraptured audience. She would have had to remain there for ever in order to perpetuate this solitude and this expectancy. But she would have had to be elsewhere as well: in the props-room, in the dressing-rooms, in the foyer; she would have had to be everywhere at the same time. She went across the proscenium and stepped up on to the stage. She opened the door to the green-room. She went on down into the yard where old stage sets lay mouldering. She alone evoked the significance of these abandoned places, of these slumbering things. She was there and they belonged to her. The world belonged to her.[12]

The smell, the half-light and the red of the carpet do not exist without Françoise, because, as appearances, their existence depends on the presence of a human consciousness. The "mission" with which Beauvoir has been "entrusted," and which she has delegated to Françoise, is to show that these appearances – in the present case those of a theatre – nevertheless do refer to a reality that exists independently of consciousness. Beauvoir's solution, and in the history of philosophy it is a breathtakingly original one, is to show that each appearance is part of a *series* of appearances, which for Françoise constitutes the independent reality or existence of the theatre. Rather than groping for the reality behind appearances, Beauvoir's analysis focuses on the serial nature of appearances, hoping to identify in such a series, properties from which may be inferred the existence of being independent of human consciousness. In the above paragraph, Beauvoir has shown three characteristics of a series of appearances. First, except for the appearance of the moment, the other members of the series are absent; it is this felt absence or lack that is the lived reality of the thing. For Françoise, it is not the isolated smell of paint or the red of the carpet or the flat of the door on which she pushes that she experiences as reality, but, rather, the partially known, but always absent, series of appearances, called the theatre, to which each moment's appearance belongs. Second, Beauvoir's paragraph shows that the unifying principle of the series of appearances constituting the theatre's reality does not depend on Françoise's whim. Beauvoir, having chosen a familiar object, invites the reader to anticipate with Françoise the series of appearances she will encounter on her walk through the theatre and, consequently, share her belief in the reality of the theatre as a being independent of her consciousness. Furthermore, the whimsical arm-stretching and calling of the empty theatre seats, inserted as it is in the otherwise objective account, calls attention to the fact that the unifying principle of the series constituting the theatre's reality does not depend on one individual's consciousness. Instead, each appearance is shown to stand in relation to *other appearances* – such as the rooms behind the doors – as well as to Françoise's consciousness. Third, "everywhere" she stands in relation to the theatre (and it would be the same for a cup or the root of a chestnut tree) offers a different appearance; so that, given the divisibility of space, "to be everywhere" relative to something is to be in an infinite number of places. The same is true with regard to changes in light. Hence, the series of appearances comprising a thing's existence is infinitely large. From these characteristics of the series it follows that consciousness can never experience all the appearances of an object and that therefore the object's being is not reducible to consciousness of it.

In an astonishingly short space, Beauvoir's analysis shows that two regions of being, consciousness and things (or in Sartre's terminology, being-for-itself and being-in-itself) arise from appearances, but that neither is reducible to the other. Even so, the philosophical content of Beauvoir's remarkable paragraph is not yet exhausted. She has also, like Sartre in his narrative encounter with the chestnut root in *Nausea*, identified individual human beings as the source of the significance of things. But her method of doing so, and her attitude or visceral response to the fact, could not be more different from Sartre's. Whereas Roquentin, *Nausea*'s "serious" and Sartre-like protagonist, is submitted to trials which cause his language to lose its grip on reality and he, himself, to fall into painful disillusionment over the contingency of meaning, Françoise joyfully, and with complete lucidity, embraces an empty theatre, a place whose *raison d'être* is the projection of transitory meanings on itself and on the world. Mouldering stage sets speak even more forcibly of the contingency of existence than the roots of a tree, and yet Françoise is completely unperturbed by their presence, and even pleased with the knowledge that it is she who evokes the significance of places and things.

From the theatre yard, Françoise passes into a garden square to observe a view of the theatre from the outside: "sleeping, except for a rosy glow from a single window." Most importantly, the square provides Beauvoir with a setting to show that, through Françoise's encounter with the deserted theatre and its abandoned stage sets, she has been addressing the same nexus of philosophical problems as was Sartre in the famous chestnut-tree passage in *Nausea*. Like Roquentin, Françoise "sat down on a bench. The sky was glossy black above the chestnut trees: she might have been in the heart of some small provincial town." Suddenly the passage becomes intensely autobiographical.

At this moment she did not in the least regret that Pierre was not beside her: there were some joys she could not know when he was with her; all the joys of solitude. They had been lost to her for eight years, and at times she almost felt a pang of regret on their account.[13]

When Beauvoir began *She Came to Stay*, it had been eight years since the oath-taking with Sartre in Carrousel Gardens. And even before that, when Sartre had turned up unannounced at Limousin, a fundamental, even metaphysical, difference between their characters manifested itself. Whereas Beauvoir was at ease with nature and environments relatively untouched by civilization, Sartre was not; whereas she experienced joy and excitement when confronted with the solitude of "slumbering things," he experienced nausea and boredom. On their very first morning together in the country, he "swept aside" Beauvoir's suggestion that they go for a walk. In her memoirs, Beauvoir notes Sartre's remarks: "He was allergic to chlorophyll, he said, and all this lush green pasturage exhausted him. The only way he could put up with it was to forget it."[14]

Of course, as has been seen, Sartre and Beauvoir compromised over this fundamental difference, and Beauvoir also pursued her lust for physical solitude and danger alone and later with Bost. But Beauvoir and Sartre's mismatched sensibilities on this point meant that their view of *things*, even from a city park bench, were contradictory, and this is reflected in both the tenor *and philosophical arguments* of their novels. Sartre's emotional fulminations against the contingency and superfluidity of things may have been more dramatic, but Beauvoir's repose in the face of the same factors enabled her to identify the philosophically cogent points of the experience, just as she had coolly picked out hand-holds on the rock faces of the mountains of the Lubéron above Marseilles. The philosophical reasoning underpinning Beauvoir's paragraph on appearances meant a great deal both to Beauvoir and Sartre, as is evidenced by the fact that Sartre used his long-winded rewording of Beauvoir's argument to begin *Being and Nothingness*. There were two reasons for her argument's importance.

Sartre, in *The Transcendence of the Ego*, had committed himself to a realist position, that is, to the idea that things exist independently of consciousness of them. But, like Beauvoir, he was equally bound to a method of philosophy that, without appeal to the supernatural, proceeded from individual consciousness. This left him with a colossal and age-old problem: consciousness reveals only the appearances of things and thus cannot show that things exist when consciousness is unaware of them. Or, at least, that was the accepted wisdom. Sartre, however, dreamed of finding a solution to the problem; and *Nausea*, which began as ann essay, was his attempt to do so. If he could found the belief that chestnut trees exist independently of consciousness of them, on the basis of reasoned argument true to his first principles, then and only then would the way be clear for him to construct the grand philosophical system of his adolescent fantasies. But this is precisely what Sartre did not succeed in doing in *Nausea*. He produced a literary masterpiece based on a powerful intuition of the reality of things, but the solid argument that could be admitted as evidence in the court of philosophers was conspicuously missing. The theory of appearances illustrated by Françoise's theatre walk had made his wildest dream possible for him.

Although Sartre's placement of "his" theory of appearances in a special chapter preceding the main body of *Being and Nothingness* is expositionally nonsensical and has caused his readers many difficulties, his motivation for doing so is easily guessed. Ever since Plato conceived of the universe as divided between appearance and reality, this division has been the Achilles' heel of philosophy. In 1912, Bertrand Russell described the problem as follows:

> Thus what we directly see and feel is merely "appearance", which we believe to be a sign of some "reality" behind. But if the reality is not what appears, have we any means of knowing whether there is any reality at all? And if so, have we any means of finding out what it is like?[15]

For centuries, but especially in the two preceding Sartre, the "great" philosophers directed their best and most courageous efforts toward closing the gap between appearance and reality; but none, suggested Russell, succeeded in developing a theory that was both credible and consistent. This failure is important because no one, least of all philosophers (as Popper bravely pointed out to the logical positivists), can pass beyond the first stages of knowledge without some opinions, manifest or otherwise, regarding the nature of reality. It is the philosopher's task to articulate and increase the internal coherence of these sets of opinions and occasionally – and these are the great moments in philosophy – to offer humankind a new transmutation of the basic ideas. Doing the latter means discovering in the common fund of possible starting points some fruitful arrangement that has escaped all the efforts of past philosophers. It is here that philosophy, against the odds, has achieved a measure of objectivity, in that, to be an initiate of one philosophical persuasion does not preclude an appreciation for the innovations of another, especially when those innovations pertain to philosophy's fundamental problems. The theory of appearances first set forth in *She Came to Stay* is just such a momentous innovation, and it was a sound career judgment on Sartre's part to place his restatement of Beauvoir's theory at the beginning of his *magnum opus*. It announced to those who know about such things that here was someone with a very special talent for philosophy and that the work that followed was to be read with the highest seriousness. Sartre had arrived on the stage of history with his own spotlight. All his far-fetched childhood dreams were now coming true with remarkable rapidity. But his principles and theories had already been thoroughly explored via Beauvoir's Françoise, who has been left outside the theatre in the dark.

> She leaned back against the hard wood of the bench. A quick step echoed on the asphalt of the pavement; a motor lorry rumbled along the avenue. There was nothing but this passing sound, the sky, the quivering foliage of the trees, and the one rose-coloured window in a black facade. There was no Françoise any longer; no one existed any longer, anywhere.

Here Beauvoir, after having considered the worldly lack of whisky and an unfinished manuscript, is reiterating Sartre's earlier point in *The Transcendence of the Ego*, that one's prereflective consciousness lacks an identity or ego. She, and later Sartre, regarded this lack as the foundation of all desire, as that which compels individuals to project themselves into the world toward chosen possibilities. Hence:

> Françoise jumped to her feet. It was strange to become a woman once more, a woman who must hurry because pressing work awaits her, with the present moment but one in her life like all the others . . . She went into Pierre's dressing-room and took the bottle of whisky from the cupboard. Then she hastened back upstairs to her office.
> "Here you are, this will put new strength into us," she said.[16]

Françoise and Gerbert need all the strength they can get because Beauvoir is about to use them to carry out the ultimate philosophical task. Through their dialogue, Beauvoir is going to do what, throughout history, 99 out of 100 philosophers have dared not do: engage directly with the question of the existence of other people as conscious beings. For 2,500 years solipsism has been Western philosophy's seleton in the closet, and only since Descartes has it occasionally been brought out for public view. The problem is that, partisanship aside, the assumptions of neither materialism nor idealism permit the deduction of the existence of other consciousnesses, nor even an analysis of the problem of how one consciousness can act on another. The awesomeness of the difficulty inherent in this problem can be appreciated by the fact that it was not until the nineteenth century that the notion of the Other, that is, a conscious being other than oneself, was first introduced to philosophy by Hegel. But he attempted neither proof of the existence of Others nor justification of knowledge of them. Even so, it was a bold step forward for Hegel to even raise the issue – too bold, because, for the next hundred years, prudence prevailed over valor as other philosophers refused to rise to the challenge of extending Hegel's very sketchy ideas on this matter. Edmund Husserl was the first one of note to do so. He, like Hegel, was mute on the question of the Other's existence, but emphasized how much human thought presupposes it. Heidegger, Husserl's student, made the Other an important part of his philosophical system, but dodged the epistemological question of the Other's existence by taking a Kantian a priori approach. Furthermore, in anticipation of his Nazism, Heidegger's ontology subjugated the individual to a mystical collective Other, rendering concrete relations between individuals unintelligible. At the end of *The Transcendence of the Ego*, Sartre attempted his own solution to the problem of other consciousnesses, touting it as "the only possible refutation of solipsism."[17] But when Beauvoir wrote *She Came to Stay*, she ignored Sartre's hypothesis and substituted her own. In fact, of the three main conclusions of Sartre's essay, Beauvoir's novel totally rejected two and radically reinterpreted the third. In *Being and Nothingness*, however, Sartre has been fully converted to Beauvoir's positions. These crucial shifts in the pair's thought need careful examination.

The truly curious thing about Sartre's essay of 1937 was that it missed the radical consequences of its principal argument. If the ego was an object of consciousness, rather than a component that distorted sense data, the need for, and even the possibility of, Husserl's technique of reduction or *epoché*, *vis-à-vis* the ego, was eliminated. If consciousness had direct access to its objects or to external reality, then the elaborate intellectual procedures of the *epoché* could be abandoned in lieu of a study of humans-in-the-world. This direct method is the hallmark of "Sartrean" existentialism; it is also the method which Beauvoir introduced and developed in her novel. But there was an even deeper implication contained in Sartre's verdict on the ego: if consciousness is sheer activity ceaselessly transcending toward objects, then it follows that reality is divided into two distinct realms *which nevertheless are inextricably linked*. These are the realms (later baptized being-for-itself and being-in-itself by Sartre) which Beauvoir so carefully delineates in the opening pages of her novel. No mention, however, is made of these foundation concepts in *The Transcendence of the Ego*. Even more interesting is their absence from Sartre's monograph *The Emotions*, whose philosophical "Introduction" was almost certainly not written before mid-1939. Here Sartre, still accepting the possibility of "putting the world in parenthesis," apologizes for the present work not being one of "pure phenomenology," and makes it clear that, for him, human and human-in-the-world are not equivalent categories.[18] But, by 1939, Beauvoir's novel, after a false start two years before, was well advanced. Comparing these contemporary texts, it is difficult not to conclude that by the middle of 1939 a large gap had opened up between Sartre's and Beauvoir's philosophical development.

The gap widens even further when considering the second and third conclusions of Sartre's essay on the ego, whose externality he elevates to a kind of moral cum methodological imperative. The fact that one's ego is merely an object of one's consciousness raises the possibility of divesting oneself of it, like Roquentin and Françoise do on their respective benches. Solipsism, argues Sartre, is to be overcome not by affirming the subjective existence of the Other, but by also throwing into doubt one's own. The cure may seem worse than the complaint, but in any case Sartre has made a serious error: with solipsism it is not an object of consciousness that is in question but rather consciousness itself. In his final conclusion he takes his confusion a step further when he declares that, given the location of the ego, "the subject–object duality . . . is purely logical" and should "definitely disappear from philosophical preoccupations."[19] In place of interacting individuals he imagines a kind of generalized Hegelian spirit, a recipe for a totalitarian society if ever there was one.

Of course, all this is terribly unSartrean, which is to say, totally opposed to the theory of Others which Beauvoir, at the end of the 1930s, was setting out in great detail in *She Came to Stay*. In its opening chapter, Françoise and Gerbert, having finished with the problem of contingency, move on to solipsism.

> "It's almost impossible to believe that other people are conscious beings, aware of their own inward feelings, as we ourselves are aware of our own," said Françoise. "To me, it's terrifying when we grasp that. We get the impression of no longer being anything but a figment of someone else's mind."[20]

It is this experiencing of oneself as another's object that Beauvoir offers as the proof of the consciousness of *others*. In the 400 pages that follow, the ramifications that this highly unstable subject–object dichotomy has for human relations are analyzed at length.

With the postulation of the Other and its ability to transform one's own consciousness, the first chapter of *She Came to Stay* has completed its sketch of the basic framework, not only of Beauvoir's philosophical universe, but also of Sartre's *Being and Nothingness*. All the philosophical matters addressed in her opening chapter, as well as additional ones, are further explored in the course of the novel. Most remarkable is the fact that her narrative, both as a whole and in its parts, is elaborately structured on the basis of her philosophical system, a fact highly significant to the question of the origin of that system and its component ideas. It shows not only that the philosophy was there in the novel from an early stage, but also that, in 1938, Beauvoir was already sufficiently fluent with its ideas to translate them comprehensively without awkwardness into a major work of fiction.

Her theory of the Other bears an especially heavy load in the novel's narrative structure, its support being crucial at three levels. Scene by scene, the relations between the characters are centered on Beauvoir's concept of the Look, which is the idea that to perceive or to imagine someone looking at oneself is to experience oneself as the other's object and hence the other as a conscious being. *She Came to Stay* persistently informs the reader who is looking at whom, a practice easily mistaken for a writer's tic but, in fact, crucial to the narrative's logic. In Beauvoir's theory there are two general modes by which two people can relate: either they can both honor each other's subjectivity, or one can play the object and the other the dominating subject. In the 1940s, Beauvoir would build an ethics based on the mode of reciprocity, but in *She Came to Stay* (as in *The Second Sex*) it was the object–subject mode of personal relations that was her primary concern. Through her characters she explores seven ways (indifference, language, love, masochism, sadism, desire, and hate) by which one can enter into an object–subject relationship. In *Being and Nothingness*, under the heading "Concrete Relations," Sartre was to offer a condensed version of the same analysis. Beauvoir was interested, too, in what happened to these dual relations when a third party (called the Third in *Being and Nothingness*) intervened, and an overlapping series of these triangles provides a third structural level to her novel.

Before leaving *She Came to Stay*, something should be said about the method Beauvoir has used in erecting the framework of her philosophy, not only because it is innovative, but also because it is the method that Sartre was to adopt in his longer and more famous statement of the same philosophical system. Beauvoir begins by observing the basic structures of consciousness: its intentionality, its prereflective self-awareness, its being-in-the-world and its union with a body, each having reference to concrete but universal experience and hence confirmable by the individual reader. She then shows how these structures of consciousness commit individuals to three ontologically primitive types of being, what Sartre was to call being-for-itself, being-in-itself and being-for-others. This procedure, basic to science but new to philosophy, of observing the structures of everyday experiences and then asking what macrostructures these microstructures entail, is the methodological bedrock of the philosophy stated in Beauvoir's novel. The propensity to shift back and forth between the concrete and the abstract, together with the commitment to consider things in-the-world, makes her philosophical method highly congenial to presentation in the novel form. When, in *Being and Nothingness*, Sartre adapted Beauvoir's method to the essay, the difference was only a change in emphasis from the concrete to the abstract, the introduction of an extensive jargon and the imposition of a great deal of rhetoric.

Suppose that instead of a man and a woman, we were concerned here with two male writers, and that, like Sartre and Beauvoir, they had each explained at length the same new philosophical system, but with one publishing their presentation before the other and gaining recognition as the system's originator. In that case, it would be quite unnecessary to write the second half of this chapter. Having established that the uncredited writer had written a comprehensive statement of the philosophy that was read by the credited one before beginning to write his own, would be sufficient to dislodge the latter from academic opinion in favor of the former as the primary thinker or "true philosopher" behind the system. Of course, this conversion of opinion would not come about overnight. Reputations and territories would have been established on the basis of the false order, and, in all probability, the less principled of their holders would fight to reobscure the facts. New books and lectures would have to be written and old ones revised. But in a decade or two – and surely it is not utopian to believe this – honor and decency among scholars would prevail with the result that the revised order of influence between the two thinkers would become accepted fact.

There is no corresponding ground for optimism in the present case. Along with most established religions, philosophy successfully continues to resist the inclusion of women in its highest echelons. Nor is this resistance limited to the philosophical establishment: the very idea of "a great woman thinker" is probably still viewed by a majority of the public – including women – as a contradiction in terms. So one must be careful not to underestimate what is at stake here: it is infinitely more than just the relative reputations of Sartre and Beauvoir; it is more, too, than just opening up philosophy's inner sanctum to women. Ultimately the question at hand is about breaking down the bigotry that underlies the contradiction noted above. And all this underlines a great irony of the present situation. Though the origin of a philosophical system is in question, its merit and importance are not. Accepted as a male creation, after half a century the system not only continues to attract scrutiny, but has also established itself as one of those extremely rare philosophical achievements whose reputation transcends the broad gulf of partisanship that separates the continental and the analytical schools of philosophy. From deep in the opposing camp, the eminent American philosopher Arthur Danto has written about the "Sartrian system," by which he means mainly the one found in *Being and Nothingness*:

> The . . . system, for its scope and ingenuity, its architectural daring and logical responsibility, its dialectical strengths and human relevance, and for the totality of its vision, is located in the same exalted category, the highest of its kind, with those of Plato and Descartes, Spinoza and Kant, Hegel and Russell, to cite most of his exiguous peers.[21]

It would be unfair to Sartre not to view his appropriation of Beauvoir's ideas in the context of the psychologically dire situation in which he found himself in the months following his departure for war on September 2, 1939. The previous summer, Beauvoir had gone on another climbing holiday in the Alps, and, in her autobiography, she gives the adventure – which lasted several weeks – a mere hundred words, saying that she "climbed every single peak between Chamonix and Tigne that was within the competence of an unaccompanied climber."[22] As is so often the case, Beauvoir's account is literally true but quite misleading. She does not mention the fact that climbing every one of those peaks with her was Jacques Bost. On July 27 she wrote to Sartre that he should know:

1. First, that I love you dearly . . .
2. You've been very sweet to write me such long letters . . .
3. Something extremely agreeable has happened to me, which I didn't at all expect when I left – I slept with Little Bost three days ago. It was I who propositioned him, of course.[23]

One more contingent coupling by either Beauvoir or Sartre was of no lasting importance to their essential relationship, but a year later, as Sartre went off to war, Beauvoir's union with Bost, although he was now also romantically linked to Olga, was threatening to rise above the contingent category. If it did, it would be the first time that either Beauvoir or Sartre had broken their oath to keep theirs their only essential relationship.

Only after the 1990 publication of Beauvoir's letters to Sartre has it become possible to gauge Bost's importance to Beauvoir. Sartre's letters to her were published in 1983, and, on their basis, biographers drew conclusions (the wrong ones, it seems) regarding the nature of the famous couple's relationship at the beginning of the war. When only Sartre's half of their war correspondence is read, one is easily led to the view that he and his interests dominated the couple's relationship. His daily letters are full of demands, not only for Beauvoir's help in his double-dealing entanglements with assorted younger women (which continued by post), but also for her to read and critique his various writing projects on which he was now, despite his official status of soldier-at-war, working twelve hours a day. In contrast, he infrequently enquires about Beauvoir's writing. But a reading of her letters to him reveals that he had no need to enquire as she kept him constantly posted of her chapter-by-chapter and draft-by-draft progress with *She Came to Stay*. She even detailed for him the hours and the cafés in which she wrote each day. And, in the months leading up to Sartre's anticipated furlough in February 1940, Beauvoir repeatedly reminded him that he would be spending part of his brief freedom on reading and discussing her novel.[24] Meanwhile, she wrote him long accounts of the various love affairs she was conducting simultaneously, sometimes describing her physical intimacies at length and frequently demanding his help in manipulating and misleading her lovers. Between their letters there is, in fact, a remarkable symmetry, but with one major exception. Reading Beauvoir's, one gradually comes to realize that, between September 1939 and February 1940, Bost was becoming for Beauvoir an *essential* love.

There is no evidence that Beauvoir ever contemplated giving up Sartre for Bost, but rather that henceforth she intended to have two essential loves rather than only one. Her letters show that, although she violated the exclusivity of her agreement with Sartre, the transparency clause remained in force. Consequently, Private Second-Class Sartre suffered a daily bombardment of illustrations of his reduced status, which began with Beauvoir's first letter after his induction. He was not yet at the front, but she feared Bost was, and told Sartre so: "The only painful thing is my intermittent bouts of panic concerning Bost: such violent pangs of dread for him that I feel I'm almost losing my reason. Especially in the evenings." In her letter of the following day, September 8, she tells Sartre how sitting at a table outside the Deux Magots "reminded me of so many things to do with you and Bost."[25] "You and Bost," and later "Bost and you," was to be the leitmotif of Beauvoir's wartime letters to Sartre.

Jacques Bost, destined to become one of France's leading journalists, was already an accomplished letter-writer, and Beauvoir placed particular emphasis on inducing Sartre to compete with him in the quantity and quality of his correspondence.[26] Thus on September 11, she opens with "no letter from you today. On the other hand, I've had another two from Bost – who's very effecting." On September 28, Sartre scores better but apparently is still not winning: "three letters from you, and three from Bost . . . His letters are all thick and very cheerful and incredibly appealing." On October 4, it is "two letters from you and two from Bost." On October 15, after speaking of "my real life – my life with Bost and you," she says, "I found two big letters from Bost – and also a clever little supplementary note he'd sent me. You too should send me an unexpected little note over and above your letters some time." But it was the scorecard dated November 1 that Sartre must have found the most wounding. "I called in at the poste restante: 5 letters from you, 6 from Bost – the clerk gave me an understanding smile. I bore my huge packet off to the Versailles, and for the first time read Bost's letters first."[27]

Even without the ever-growing threat of partial or total betrayal by Beauvoir, Sartre was living through an enormously difficult period. For only the second time in his life – the other being his childhood ordeal in La Rochelle – he was condemned, like other mortals, to exist indefinitely without the support of a daily chorus of adulation. Except in the disembodied form of letters, which he now feverishly sought, his life was devoid of the intellectual conversations in which he shined, of the caresses of his youthful girlfriends, of the motherliness and camaraderie of Beauvoir, of the uncritical admiration of his students, and of the increasing celebrity that had accompanied him in his Parisian café life. In their place was a straw mattress which he shared with another Private Second-Class, a tin helmet which, like his fatigues, was too big for him, and the constant living at close quarters with men whom he described as "big guys who shit, wash themselves, snore, and smell of man."[28] Worse still, his immediate superior, Corporal Pierre, was an authoritarian mathematics teacher who must have reminded him of his hated stepfather Mancy.[29] Sartre's defense was to withdraw from the society of his fellow soldiers into a frenzy of writing, broken only by the need to sleep, eat and launch weather balloons twice a day. In addition to his daily output of several long letters and five pages of his novel, he kept a diary that in nine months ran to fourteen notebooks. This journal, whose five surviving notebooks were published posthumously as his *War Diaries*, became for him what he called his "secret life."[30] Meanwhile the pathological nature of his behavior did not go unnoticed by his fellow draftees. In the French National Archives, Cohen-Solal has discovered how Sartre was perceived by the soldiers in his group. Sartre, she writes, was famous among his fellow conscripts

> for the weeks he spends "without taking a bath when all he had to do was cross the street and pay ten sous to have exclusive use of a bathroom in the heated building," and for his nickname in the barracks, "the man with the black gloves," because his hands "were black with dirt up to his elbows." We also have a number of anecdotes, such as this one, provided by the corporal: "Once I had a real fight with him while trying to stop him from burning the furniture of the house we lived in: he didn't want to waste time chopping wood when he could read or write instead."[31]

It is perhaps significant that Sartre's diary-keeping was a reversion, both in motivation and literary form, to the patterns of his early childhood writing. In writing his childish "novel notebooks," he says, he had been "escaping from the grownups," and, in writing his diary, it is obvious he was escaping from the ever-present soldiers. At age 9 he had "indiscriminately poured everything [he] read, good or bad," including "odds and ends of gloomy tales and cheery adventures, of fantastic events and encyclopedia articles," into his notebooks; at age 34 he did much the same in his diary.[32] He philosophized about the books Beauvoir sent to him (and to Bost), chronicled the absurdities and squabbles of his army life, copied out letters from friends, sketched portraits – invariably disapproving – of his associates, and shamelessly retold their stories. His childhood notebooks featured a swashbuckling hero with whom Sartre identified, but in his war notebooks Sartre has grown up to play the hero in his own name, and it is his quest to reform himself by self-examination into an ever more remarkable person and to construct an original philosophical system that could save the world that provides a plot for his otherwise obsessional writing. One of several surprising things about the *War Diaries* is that, on the whole, it is a remarkably good read. Its prose has a fresh, rough-hewn, camera-eye quality; and Sartre, who in writing about himself has not edited out the many passages that give him away as a poseur, is obsessed with how the world – past, present and, most of all, the future – might perceive him. He was not yet in any mortal danger, but at any moment that could change, and if all-out war between France and Germany began, then there was the very real possibility that, with his diaries, he was writing his last work. On September 16, he wrote to Beauvoir telling her that if he died, she was to publish his diaries and add "benevolent and explanatory annotations."[33] Meanwhile, in what was clearly either a last-ditch attempt to secure immortality as a philosopher in the event that his life should be cut short, or a repository of ideas for a major work if he should live, he tossed in among his descriptions of barracks life and childhood remembrances every seemingly novel philosophical thought that passed through his head. But when, in January 1940, he read through what he had written, he realized that his quest for philosophical originality had been a dismal failure. On January 9 he wrote to Beauvoir about his disillusionment.

> I have reread my five notebooks, and they don't please me nearly as much as I had expected. I find them a little vague, too discreet, even the clearest ideas are little more than rehashings of Heidegger's: in the end, all I have done since September . . . is only a long re-elaboration of the ten pages he devoted to the question of historicity.[34]

Sartre's fortunes changed dramatically on the morning of February 5, 1940, in a Paris brasserie on the avenue du Maine. It was there that Sartre discovered the mother lode of philosophical ideas which would win him worldwide fame as a philosopher. In her journal, Beauvoir recorded how, as she read his diaries, he began to read *She Came to Stay*. He was at the beginning of ten days of leave, and in the first five he had seven sessions of reading Beauvoir's novel in her presence. On his last day before departing, Beauvoir's journal says that Sartre read the last chapter of her novel, which she had rewritten during his stay.[35] When, on February 16, Sartre returned to camp and his diary, he found himself, for the first (and only) time in his life, overflowing with original and major philosophical ideas. Sensing that there were more to be extracted from the lode back in Paris, Sartre's dream of becoming *known* as one of history's greatest philosophers suddenly must have looked to him like a real possibility.

On February 17, his first full day back at his army post, Sartre began trying to put down in his diary the philosophical ideas he had unearthed in Paris. At first, it appeared that his effort would be an honorable one. He writes as follows.

> For the Beaver has taught me something new: in her novel, one sees Elisabeth complaining about being surrounded by objects she'd like to enjoy, but that she can't "realize" . . . She meant that we are surrounded by *unrealizables*. These are existing objects, that we can think from afar and describe, but never *see*.

For a thousand words Sartre continues to elaborate on Beauvoir's concept. Unfortunately, he fundamentally misremembered it: Beauvoir's system requires that the concept of unrealizables has to do with situations, not objects. By the next day, Sartre understood this, and he corrects himself in his diary: "My comment yesterday about unrealizables could give rise to confusion. What is unrealizable is never an *object*. It is a *situation*. It's not Paris, but being-in-Paris, with respect to which the question of the unrealizable is posed."[36]

In some ways, these entries on *unrealizables* are representative of the method Sartre used to understand and transcribe parts of Beauvoir's philosophical system during the first eleven days after his leave. He takes up one element at a time, invents the terminology, and then, over several days, works toward an approximation of the underlying theory which he had briefly encountered in Beauvoir's text. However, *unrealizables* is only a minor concept in her system; it is also the only one for which Sartre in his diaries gave her credit. (When he came to write *Being and Nothingness*, even this small act of honesty was withdrawn.)[37] On February 18, 1940, having thrown his crumb of credit to Beauvoir, he moved on to laying claim to major elements of her philosophy. As the philosopher-hero of his diaries, Sartre had been slow to justify his opinion of himself through the delineation of any important new ideas. Now that he was about to articulate the philosophical ideas that would allow him to believe in his own success, he savors the moment when his childhood reveries were about to turn into something more substantial than dreams:

> I feel strangely bashful about embarking on a study of temporality. Time has always struck me as a philosophical headache, and I've inadvertently

gone in for a philosophy of the instant (which Koyré reproached me for one evening in June '39) – as a result of not understanding duration . . . I was extremely embarrassed and put out to see myself as the sole instantaneist cast among contemporary philosophies which are all philosophies of time. I tried in *La Psyché* [a mostly unpublished work] to derive time dialectically from freedom. For me, it was a bold gesture. But all that wasn't yet ripe. And, behold, I now glimpse a theory of time! I feel intimidated before expounding it, I feel like a kid.[38]

With Bost now on leave and usurping Sartre's place with Beauvoir in Paris, it is impossible not to be reminded here of Sartre's thefts from his mother's purse in La Rochelle. The adult Sartre hopes to impress his readers with the theory of time he has glimpsed in Beauvoir's manuscript. In her novel, Beauvoir develops her theory of time incrementally, but early in the text she has Pierre, the Sartre-like character, state the bare bones of her theory. Pierre says:

". . . but it's also impossible to live only for the moment."

"Why?" said Xavière. "Why do people always have to drag so much dead weight about with them?"

"Look," said Pierre, "time isn't made up of a heap of little separate bits into which you can shut yourself up in turn. When you think you're living purely in the present, you're involving your future whether you like it or not."

"I don't understand," said Xavière . . .

"Let's assume you've decided to go to a concert," said Pierre. "Just as you're about to set out, the idea of walking or taking the metro there strikes you as unbearable. So you convince yourself that you are free as regards your previous decision, and you stay at home. That's all very well, but when ten minutes later you find yourself sitting in an armchair, bored stiff, you are no longer in the least free. You're simply suffering the consequences of your own act."[39]

Here, and at other points in her novel, Beauvoir is expounding, not a theory of physical time, but rather a theory of how time is experienced by human beings, that is, of temporality. It also must be emphasized that she is not moralizing, in the manner of Heidegger, about how one *should* deal with time. While her theory bears a superficial resemblance to the German's, it exists on a fundamentally different level, and Beauvoir has taken pains to drive this point home by beginning her exposition with the *assertion* that it is *impossible* to live only for the moment, in contradistinction to Heidegger's injunction that one *should* not do so.

Clearly, Beauvoir's crediting of this piece of philosophical wisdom to the fictional Sartre, Pierre, was more than the real Sartre could resist. But his first attempt in the *War Diaries* at reconstituting Beauvoir's theory of temporality, despite his fanfare introduction, degenerated quickly into pompous gibberish ("Time is the opaque limit of consciousness. It is, moreover, an indiscernible opacity in a total translucidity"), before recovering briefly with: "time appears to us only thanks to the *past* or the *future*: it is not given us to live in its continual flow." The next day, however, his memory served him well as he produced the following near-perfect jargonized generalization of Pierre's speech to Xavière about deciding to stay at home.

> And since this immediate past is negation of a more distant past, and so on and so forth, it is by this nihilation of the total bloc of the past it *has been* that the present for-itself defines, in its presence. Thus the question cannot arise of knowing why freedom cannot escape this past, or give us another past – since, precisely, we are free *with respect* to this past.[40]

In *She Came to Stay*, Beauvoir argues that the future, no less than the past, is inextricably tied up with the present.

> The day had been spent in the expectation of these hours, and now they were crumbling away, becoming, in their turn, another period of expectancy . . . It was a journey without end, leading to an indefinite future, eternally shifting just as she was reaching the present.

Repeatedly, Beauvoir shows characters overcome with anxiety about their "journey without end," who try to escape their future by denying their consciousness of the world. Sartre would have read her following example.

> If only she could fall asleep again till tomorrow – not to have to make any decisions – not to have to think. How long could she remain plunged in this merciful torpor? Make believe I'm dead – make believe I'm floating – but already it was an effort to narrow her eyes and see nothing at all.[41]

To remain listless, contends Beauvoir, is no less a projection of oneself into the future than deciding to cross town to go to a concert, and, whatever decision is made – even the decision not to decide – brings consequences (for example, of being bored, or having or not having a university degree), which become part of one's present and future. *Thus past, present and future all stand in the relation of co-implication*; and conversely, the present is not even conceivable without the past and the future. The beauty of all this, which Beauvoir exploits to the utmost in her novel, is that this theory of temporality dovetails perfectly with her theory that consciousness is an emptiness (or nothingness) that ceaselessly requires filling and that, except through sleep or death, one cannot escape, projecting oneself continuously toward the world. In this existential emptiness or lack, Beauvoir succeeds in grounding all the desires of her characters and, by implication, human desire in general. When Elisabeth "looks into herself," says Beauvoir, "all she finds is an empty shell. She had no idea that it's the common fate."[42]

On February 22, 1940, having exhausted his stock of Beauvoirian wisdom on the past, Sartre, in his diary, moved on to "the nature of the future" and, in particular, the way in which Beauvoir linked it to the nature of consciousness. "The Future," wrote Sartre, "could exist only as complement of a lack in the present. It is the very signification of this lack." Sartre then goes on to show that he recognized a revolutionary idea when he read one.

> It's quite astonishing that, in all philosophies and in all psychologies, it should have been possible to describe at length will, desire or passion, without being led to see the essential fact: namely, that none of these phenomena can ever be conceived if the being which wills, suffers or desires is not gripped in its being as afflicted by an existential lack.

He also saw how this element of Beauvoir's philosophical system undermined that of her chief rival, Heidegger. After crediting Christianity with having come closest to recognizing that the human soul is "animated" by an emptiness, Sartre continues:

> Yet it must be noted that most Christian thinkers, led astray by their monist conception of being as an *in-itself*, have confused – like Heidegger, moreover – the existential nothingness of human consciousness with its finitude. Now finitude, being an external limit of being, cannot be at the root of lack, which is found at the very heart of consciousness.[43]

Sartre turned his attention from temporality and desire to the existence of the Other on February 27. In December he had written, "Love is the effort of human reality to be a foundation of itself in the Other."[44] In *She Came to Stay*, Beauvoir repeatedly makes the same point, as when she says that, in loving Pierre, Françoise was "looking upon him only as a justification of herself."[45] Sartre and Beauvoir's observation about love is a worthy one, but in and of itself it is neither original nor of philosophical interest. In Beauvoir's novel, however, this pedestrian observation acquires deep philosophical resonance when it is placed in the context of a general theory of human existence *vis-à-vis* the Other. Just as Beauvoir's description of Françoise and Pierre's relationship appears as an exemplification of her analysis of love, so the latter appears as an exemplification of Beauvoir's more general theory of human relations.

The same is true of the novel's investigations of indifference, language, masochism, sadism, desire, and hatred, so that they all, including all the details of the actual behavior of the characters, refer analytically back to each other. It is this which, in large part, accounts for the peerless philosophical unity of Beauvoir's narrative, a unity of which Sartre, as reader, would have been keenly aware, and whose basis he tried to ferret out in his diary entry of February 27. He made some progress. One exists, he wrote, "as a defenceless object projected onto the Other's infinite freedom," and one's "only way of *not-being* the Other *is* to *be-for* the Other."[46] In a fuzzy way, he has identified the nature of the subject–object duality that underpins Beauvoir's work. But, despite going on for pages and repeated fresh starts, he fails to reproduce Beauvoir's concepts of the Look and of the Third. His discussion of concrete relations is desultory and limited mainly to love and sadism, and, astonishingly for someone who thought of himself as a philosopher, he makes no mention of solipsism. Clearly, Sartre needed more reading time with *She Came to Stay* and more tutorials with Beauvoir before being able to write the brilliant exposition of her theory of being-for-others that would appear in *Being and Nothingness*.

Fairness to Sartre requires mention of two more factors that may partially excuse his initial unacknowledged thefts from Beauvoir's philosophy. On February 20, he received Beauvoir's letter of February 18 in which she formally announced that henceforth her relationship with Bost would be an *essential* one. Though Beauvoir had scarcely kept it a secret that things were moving in this direction, receiving her announcement must have been painful for Sartre. Bost was now with Beauvoir in Paris and, for the first two pages of her letter, she only hinted at what needed to be said: "and though we talk non-stop from morning to night we'll never get to the end of what we have to say to one another." In the end she put her position to Sartre straight:

> There's one thing of which I'm now sure, which is that Bost forms part of my future in an absolutely certain – even essential – way. I felt such "remorse" because of him, that I want a postwar existence with him – and partly *for* him.[47]

Sartre's comments in old age suggest how he may have felt in the period following Beauvoir's declaration.

> Basically I didn't much care whether there was another man in an affair with any given woman. The essential was that I should come first. But the idea of a triangle in which there was me and another better-established man – that was a situation I couldn't bear.[48]

What is certain is that, after his own letter to Beauvoir of February 20, in which he put on a brave face of indifference to Beauvoir's announcement, Sartre's behavior in relation to her, even ignoring his diaries, became quite extraordinary. His instinct seems to have been to retaliate, but he was in a weak position to do so. His girlfriends, like Wanda (who had finally slept with him in the previous summer), had been chosen for their sexual charms rather than for their intellectual claims; none were plausible candidates for an essential relationship with him in the way that Bost was for Beauvoir.[49] Even so, it was inevitable that Wanda, with whom he had spent time while on leave, would now take on added value in his eyes (even though Sartre was willing to describe her as having "the mental faculties of a dragonfly"), and he sought to cement their shaky relationship.[50] She, most likely sensing an opportunity, protested in a letter at the "mysticism" of his love for Beauvoir. Sartre wrote back: "You know very well that I'd trample on everybody (even the Beaver despite my 'mysticism') to be on good terms with you." Or at least, in a letter to Beauvoir on February 24, this is what Sartre said he had said. His words seem to have been aimed more at wounding Beauvoir than at placating Wanda. And it also seems that they were intended to wound himself, because immediately after writing them, his letter became abject with self-condemnation: "I feel profoundly and sincerely that I'm a bastard. And a small-scale bastard on top of that, a kind of university sadist and a sick-making civil service Don Juan. It is necessary to change this." For the next week Sartre continued to write to Beauvoir in this confessional, self-abasing vein, and with shrill promises of "no more affairs for a long time," promises for which Beauvoir, in her letters, was *not* asking.[51]

Philosophical systems like the one found in *She Came to Stay* are not developed overnight nor even in a year. Yet, as is shown by Sartre's writing of 1939 and 1940, prior to his first army leave, Beauvoir had kept him almost totally in the dark concerning her philosophical theories. Indeed, she appears to have carried out the whole process of their development in secret; and this, surely, was a major violation of the couple's vow never to "conceal anything from the other."[52] Sartre must have seen this as a further betrayal by Beauvoir; perhaps this also mitigated his guilt for appropriating as his own the philosophy that must have been created largely without him.

The foregoing discoveries regarding the origins of Sartrean existentialism throw up a series of intriguing questions. Why did Beauvoir let Sartre get away with it? Why did Sartre not destroy his war diaries? What, if any, was the arrangement between him and Beauvoir? Did she intend for the truth to be uncovered after her death? Except for the last question, there is little to offer to date but speculation.

Understanding Beauvoir the philosopher may be helped by considering an early event in Sartre's life. When his father died, his mother whisked him straight from the burial to the train station. Anne Marie fled north with unseemly haste because, if she remained near her male in-laws, she risked losing possession of her child. Her worthiness as a mother was not in question; but the Sartre men were a queer lot, and being a woman meant that she had no legal entitlement to the outright possession of her own child, indeed of anything, except perhaps the clothes on her back. A generation later, social morals and legal reform had changed women's position, but perhaps not nearly so much as one might think. In France, in 1938, women still had not won the vote, and, more to the point, nowhere did their right to the proprietorship of ideas exist as more than an improbable and eccentric dream. It was in this situation that Beauvoir decided what to do with her philosophical system.

The publisher's rejection of her short story collection (*When Things of the Spirit Come First*) in 1937, besides being traumatic, gave Beauvoir a valuable object lesson. Gallimard, who had published Sartre's *Nausea*, had rejected her much tamer work explicitly on the grounds that it violated society's rules as to what kind of material could be published by and about women, and told her that a particularly "subversive" aspect of her collection was that it reflected "what women think."[53] So, from 1937 onwards, Beauvoir knew that the criteria for publication she faced in France were radically more restrictive than those faced by Sartre, and that her best chance for becoming an author was to make her future work comply, or at least appear to comply, with the special set of rules enforced by Gallimard and others against her sex. Even today, as Sylvia Lawson notes, women are more restricted in what is regarded as fitting topics for them than it might appear. Despite the gains the present feminist movement has brought, women continue to be plagued by narrow expectations regarding their interests, and the intellectual woman who challenges publishers by stepping "outside what's become a comfortable feminist shopping-list of issues, is an irritant and a troublemaker still."[54] Half a century ago Beauvoir's position was immeasurably more difficult. But this did not mean that, under the social conditions of her day, Beauvoir could not reasonably hope to have a philosophical text published. On the contrary, she could write an entire book on what a woman thought about philosophy if she disguised it as something else. Sartre had written and published an important philosophical text in the form of a novel; she could do the same if she did not call attention to her philosophical content.

There are strong and repeated suggestions in *She Came to Stay* that Beauvoir both intended and expected Sartre to take up her philosophy as if it were his own. In Chapter 1, as Beauvoir outlines her new version of Western philosophy, Françoise is creating a new version of *Julius Caesar* which will be identified as Pierre's production and in which he will star as Caesar. Later, attention is drawn to the fact that, in contrast to Pierre, Françoise and her typist, Gerbert, will not receive credit for their efforts; Françoise says, "We like to do our work well; but not for any honour or glory." At the play's first performance, when Pierre begins to deliver his lines, Françoise

> felt as if they sprang from her own will. And yet it was outside her, on the stage, that they materialized. It was agonizing. She would feel herself responsible for the slightest failure and she couldn't raise a finger to prevent it.
>
> "It's true that we are really one," she thought with a burst of love. Pierre was speaking, his hand was raised, but his gesture, his tone, were as much a part of Françoise's life as of his.[55]

No matter how much Beauvoir loved and admired Sartre, no matter how well she anticipated precisely his moral failings, and no matter how resigned she was to the limitations society imposed on her as a woman, it must have been "agonizing," indeed, for her to listen quietly to Sartre as he flourished her ideas, sometimes muddling them, and always claiming them as his own.

Sartre's integrity in all this, even without the entries in his *War Diaries*, seems irredeemable. By the early 1940s, his reputation was already so well established that, if he had wished, he could have arranged to have a woman's name appear on the cover of *Being and Nothingness* along with his own. Other scenarios whereby Sartre could have gradually unveiled to the world the culturally traumatic fact that one of its greatest philosophers was a woman are easily imaginable. To be sure, once the true origins of "Sartrean existentialism" were disclosed, it might no longer have been taken with the same seriousness, but it also might have been the case for Sartre to choose truth and a contribution to the fight for equality over the consolidation of his own reputation.

But care must be taken not to exaggerate Sartre's crime. In his formative years, Sartre was encouraged in certain proclivities – plagiarism, egomaniacal ambition, and common, everyday, unreconstructed male chauvinism – which together made him highly susceptible to the temptations posed by the delights forthcoming from Beauvoir's mind. And, certainly, Sartre was encouraged in these weaknesses by Beauvoir. Her puppeteering of him and the huge deception she successfully foisted on the world also duplicated, on a grander scale, the patterns of her personal life and so must have been enormously satisfying to her own unusual appetites. Her public reputation did not come at the top of her priorities, and in any case, no matter how many ideas she gave away to Sartre, in the end there would be enough left over for her to become famous too.

Then there is the matter of Sartre's relation to Sartre. Certainly his obsessive interest in Jean Genet's life as a thief is no longer particularly puzzling. But the question remains as to whether Sartre himself was acting in bad faith about the origins of Sartrean existentialism, or did he convince himself, as he convinced the world, that he was its originator? When interviewed with Beauvoir, he often showed uncharacteristic modesty about his philosophical achievements, suggesting that, in her presence, he was, on some level, aware of his secret debt and of possible limits to Beauvoir's forbearance. But if one does not believe that Sartre wanted the truth to be discovered after he died, then his failure to destroy the incriminating entries in his war diaries is compelling evidence that he was, in the main, a "victim" of bad faith.

In his maturity Sartre "adopted" two younger women as his daughters. With one, Arlette Elkaïm, he went through legal adoption procedures so that, as he explained to John Gerassi, there would be someone after Beauvoir died to keep distributing his royalties to the five women whom he financially supported, including Elkaïm herself.[56] But, following Sartre's death, events took an unexpected turn when Elkaïm, whose legal rights prevailed over Beauvoir's moral ones, took it upon herself to decide which of Sartre's unpublished works to release. Consequently, Beauvoir, who was scarcely on speaking terms with Elkaïm, was powerless to stop the publication of Sartre's *War Diaries* in 1983. Their appearance threatened the Sartre–Beauvoir legend, which Beauvoir (although arranging for its unravelling after her death), was desperate to see preserved during her lifetime. The danger from *War Diaries* was threefold. First, Sartre's entry of February 17, 1940, where he credits Beauvoir with the concept of unrealizables, shows that he read *She Came to Stay* as a *philosophical text*. (It was this that led the present authors to do the same.) Second, the conspicuous cluster of original philosophical ideas in Sartre's diaries immediately after his February leave invited scholars to look for special input during his Paris fortnight, and his diaries would tell them that he had read Beauvoir's novel. So, in this way, too, the *War Diaries* could have led to a decoding of *She Came to Stay*. And third, once the decoding had been effected, comparison of the philosophical contents of the novel to Sartre's diaries would reveal who really had been leading whom and, moreover, that, in writing *Being and Nothingness*, Sartre had been acting primarily in the role of interpreter and elaborator for the philosopher, Simone de Beauvoir.

From 1983 on, Beauvoir apparently regarded the discovery of the philosophical system buried in *She Came to Stay* as imminent, and, against this eventuality, concocted for her biographer, Deirdre Bair, an elaborate series of stories to the effect that she had scarcely begun the novel at the time of Sartre's February leave. Bair, though philosophy was not her strong suit, may have spotted Sartre's February 17 entry and confronted Beauvoir directly with what it suggested about his relation to her novel. What is certain is that Bair's lengthy description of Beauvoir's oral account of the writing of *She Came to Stay* is centered on Sartre's February leave. It is also clear that, in Beauvoir's mind, the crucial matter in question was which part of her novel people were to believe she had written before and after his stay in Paris. Previously, in *The Prime of Life*, Beauvoir wrote at length about the conception and writing of *She Came to Stay*, but that did not keep her from fabricating a radically different story to fend off the combined threat posed by Bair and Sartre's *War Diaries*.[57] Beauvoir's earlier account, which is consistent with the facts revealed in her letters and journals, tells how, after making a false start on the novel in October 1937, the following autumn she began it again with a new plan which included the relationships between the five major characters and everything from the appearance of Gerbert and Françoise in the first chapter to the killing of Xavière in the last.

That Beauvoir persuaded Bair that her previous account was all lies and that the new one was the truth can probably be attributed to the force of Beauvoir's personality. In any case, Bair wrote: "Actually, by February 1940 she had written what would later amount to less than fifty pages of the printed text, nor did she fully conceptualize it until Sartre's first leave, in that month."[58] Beauvoir's task was complicated by the fact that the scene concerning Elisabeth and the concept of unrealizables discussed in Sartre's February 17 diary entry appears early in the second half of the novel. However, Beauvoir persuaded Bair to write that "by the time of Sartre's February leave, she was no further along with the novel than the initial conception and the development of" Françoise and Elisabeth, and that the only parts of her fifty pages that she "wanted Sartre to read during his leave did not concern her main character, Françoise, but rather were about Elisabeth . . . ."[59] Since very little of the philosophical content of *She Came to Stay* is expressed through Elisabeth, this farrago of untruths, if believed, would eliminate Beauvoir's novel as the source of the philosophical ideas that appeared in Sartre's diaries immediately after returning from his February leave.

In 1983 Beauvoir published Sartre's letters, her timing perhaps inspired by hope of diverting attention from Sartre's *War Diaries*. But the letters' publication seems to have increased the awkwardness of the questions put to her by Bair, who was midway through the interviews she conducted with Beauvoir over the last six years of her life. Why, Bair asked Beauvoir, did she not publish her letters to Sartre or at least let her biographer read them? At first Beauvoir denied their existence; later, in the face of Bair's incredulity, she admitted that she did have them but that they were "not interesting." But Bair was very interested. "On several occasions," wrote Bair, "our next meeting began with a gruff Beauvoir waving a sheaf of her letters that she had supposedly 'just found in the cellar among other things', all of which supported" her claim that they contained "nothing of any value."[60]

To her adopted daughter, Sylvie Le Bon, Beauvoir insisted to the end that her own letters to Sartre were lost. But in a 1984 interview with a Canadian feminist magazine, Beauvoir categorically stated: "I don't feel I ought to publish letters of my own during my lifetime. When I'm dead they might perhaps be published, if they can be found." Sylvie did find them.

> One gloomy day in November 1986, while rummaging aimlessly in the depths of a cupboard at her place, I unearthed a massive packet: letters upon letters in her hand, most of them still folded in their envelopes. Addressed to "Monsieur Sartre". It was as unexpected and moving as suddenly discovering a secret chamber in a pyramid explored countless times. She had been mistaken, her letters did exist.[61]

In fact, Beauvoir had left very little to chance. Beauvoir's "place" was only a small studio apartment, and she knew that Sylvie, when she found them, would publish her letters which would prove to the world that the professors at the Sorbonne had been more right than they could ever have imagined when they judged that, of the pair, Beauvoir, not Sartre, was the true philosopher.

6

THE WAR YEARS

Whatever the state of the evolution of their philosophical, literary and sexual negotiations, when the nine months of the Phoney War ended in early May of 1940 and the German invaders swept through Holland, Belgium and Luxemburg to gain control of France after only six weeks of fighting, Beauvoir and Sartre's world was changed forever. They both saw the war as a shocking turning point in their attitudes and preconceptions about the world. Before it, they had, somehow, felt themselves exempt from the general fortunes of their society, in relation to which they prided themselves on their positions as intellectuals, bohemians, and rebels against the proprieties. After the war, both Sartre and Beauvoir demonstrated, in various ways, their revised awareness of the need for writers to be politically engaged. In personal terms, they discovered themselves as creatures of history.

If, during the Phoney War, neither Sartre nor Beauvoir could quite bring the potential conflict into focus as anything but a rather tiresome interruption of their authentic existence, both later identified it as a monumental fact that turned their lives in new directions.[1] As Beauvoir says in *The Prime of Life*, dividing lives into significant sections is always an arbitrary process – though the couple was indisputably fond of trying to do so – but the impact of the Second World War, coming, as it did, so soon after the triumph of Sartre's real breakthrough into literary success, and their joint excitement at fashioning the grounds of a new philosophical system, stunned them both.[2] Beauvoir says that, while there is no way that she could name the day or the hour, sometime in 1939 her life reached a watershed. She said, "I renounced my individualistic, anti-humanist way of life. I learned the value of solidarity." Prior to the cataclysm,

128

she had only "two preoccupations: to live fully, and to achieve my still theoretical vocation as a writer."[3] Her egoism, she said, was still intact. The relationship with Sartre, though its sexual ardor had faded and Beauvoir's attachment to Bost had complicated it, was a further *égoisme à deux*. As she saw it, they had "pioneered" the nature of their association, "its freedom, intimacy, and frankness." And while "on an intellectual plane" they regarded themselves as "both honest and conscientious," with a "genuine sense of the truth," they were nevertheless protected, in ways they had not fully registered, from most forms of hardship. The war made them see what their position was. As teachers, with posts protected by comfortable conditions of employment as servants of the state, the Depression and its suffering had made little impression on their lives and less on their imaginations. The Spanish Civil War had interested them only mildly and in an abstract way, despite its impact on their friends, the Gerassis. As Beauvoir said, up until Sartre's mobilization in 1939, the pair had felt themselves immune from most kinds of accountability: "Like every bourgeois, we were sheltered from want; like every civil servant, we were guaranteed against insecurity. Further, we had no children, no families, no responsibilities: we were like elves."[4]

"The fact of the world's adversity," said Beauvoir, was concealed from them by the structural freedom granted by their status. When the war came, recalled Beauvoir, it knocked her individualism to pieces: "suddenly History burst over me, and I dissolved into fragments. I woke to find myself scattered over the four quarters of the globe, linked by every nerve in me to each and every other individual."[5] While Beauvoir scarcely gave up her pursuit of individual happiness, even during the war, this goal now took a more problematic place on her scale of values.

For Sartre, the war confirmed his personal opening out to contingency that had ended his period of mental instability which had lasted from 1935 to 1937.[6] Before the conflict, he said, he thought he was "leading the life of a totally free individual." "From '39 on" he realized that he "no longer belonged to himself." During the war, he said, he "began to reflect on what it meant to be historical, to be part of a piece of history that was continually being decided by collective occurrences."[7] Nothing could be further from the "great man theory of history" that he had taken in with his grandfather's encouragement in his childhood.

As the Panzer divisions rolled across France, and after the débâcle of Dunkirk on May 31, 1940, neither Sartre nor Beauvoir had the opportunity to consider themselves as special cases. The Battle of Paris began on June 5. Beauvoir, who now knew absolutely that "life had finally ceased to adapt itself to my will," followed the advice Sartre had given her in May.[8] She packed the Kosakiewicz sisters off to the country, then, on June 10, she joined the six million of her fellow countrymen and women who took to the roads in a panicked flight before the invading German troops. Concerned about the details of her teaching post until the last minute, and sickened at the thought of being irrevocably divided from Sartre (and Bost), Beauvoir left Paris with the help of Bianca Bienenfield, whose father's car took her south on her journey to the safety of Madame Morel's La Pouëze near Angers. Beauvoir lost her luggage during the trip, a nightmare of confusion and terror as millions of fleeing refugees clogged the roads of France. Marshal Pétain, the 84-year-old head of the French High Command, announced his country's surrender on June 17. The Occupation had begun. As James D. Wilkinson points out, for "the next four years, France experienced World War II not as a belligerent but as a captive." As one writer put it, for the time, "all of France, all of Europe is in prison."[9]

On June 21, the terms of the French armistice were made public: on June 28, Beauvoir, bored, despairing, yet somehow convinced that Sartre might have returned to the capital, set off on a difficult journey back to Paris. When she got there, she discovered that her landlady at the hotel on the rue Vavin had thrown away all her belongings. She also ran into her father who told her that the POWs were interned in "vast camps" where "they were starving to death on a diet of 'dead dog'."[10] Profoundly depressed and wondering why she "so absurdly continued to survive," frightened at the thought of Sartre "literally and physically starving to death," Beauvoir picked up her life as a citizen of a city under foreign occupation, and waited to hear what had become of her friends and lovers.

On July 11, the day after Pétain had been installed as the new Prime Minister of the Vichy government, Beauvoir received a letter from Sartre which had been written on July 2. He was held prisoner in a camp near Baccarat; he said he was not badly treated; he was working a little; could she send food.[11] On June 30, Beauvoir had recorded her belief that there would be an "afterwards" to the war. With Sartre's letter in her hands she felt she breathed "a little easier." Her personal defeat would be one more triumph for the Germans; she determined, as best she could, to "become a *person* again, with a past and future" of her own.[12] She learned to ride the bicycle that her student and lover, Nathalie Sorokine, stole for her, read Hegel in the Bibliothèque Nationale, and got on with her life while she waited for Sartre to return.

If Beauvoir's way of adjusting to the captivity of France was to insist stubbornly on leading her life in as normal a way as possible, Sartre's experience was altogether more traumatic. When he was taken prisoner, on June 21, 1940 – his thirty-fifth birthday – he said that he learned in a stroke "what historical truth really was."

Aside from what they reveal about the couple's philosophical development and the adjustments in their relationship, the letters that Sartre and Beauvoir exchanged during the Phoney War from September 1939 to June 1940 are testaments, not only to their continuing need to reflect their experience to each other, but to their sense of the unreality of the entire situation. During this time there was no fighting on the French front, and Sartre merely continued to perform his absurd meteorological exercises in a vacuum that he found ridiculous. A letter from Sartre, written during his first army posting at Brumath, which Beauvoir transcribed in *The Prime of Life*, demonstrates his view of the stupidity of his position:

> My work here consists of sending up balloons and then watching them through a pair of field glasses: this is called "making a meteorological observation." Afterwards I phone the battery artillery officers and tell them the wind direction: what they do with this information is their affair. The young ones make some use of intelligence reports; the old school just shove them straight in the wastepaper basket. Since there isn't any shooting either course is equally effective. It's extremely peaceful work (I can't think of any branch of the services that has a quieter, more poetic job, apart from the pigeon breeders, that is, always supposing there are any of them left nowadays) and I'm left with a large amount of spare time, which I'm using to finish my novel.[13]

Sartre found not only his work but his fellow conscripts ridiculous, despite his resolutions to become friendly with them. It was, as has been noted, his first chance to live among men who were not protected by educational privilege, and he seems to have made little of it. Indeed, Sartre spent most of the Phoney War writing. He composed at least three letters a day (to Beauvoir, to Wanda, and to his mother), and wrote an average of five pages of his novel, *The Age of Reason*, as well as four pages of the journal that would be posthumously published as *The War Diaries*.[14] His and Beauvoir's letters during this time represent their attempt to hold onto their sanity in a world gone mad. Beauvoir's letters to Sartre attempt, at his request, to keep an unbroken picture of the life of the capital alive in his mind. These letters, in addition to conveying to Sartre the changing state of Beauvoir's feelings for Bost, and the details of her progress with her novel, are full of tantalizing trivia, gossip, details of the most ordinary aspects of her life – the food she eats, the cafés she frequents, the times at which she goes to bed – and outrageous reports on the appearance (and sometimes the smell) of her lovers. They are exercises meant to show that the ordinariness of life keeps functioning, no matter how strained the circumstances. Sartre's letters to Beauvoir are also records of everyday reality, addressed to the woman he now said he regarded as "my conscience and my witness."[15] Despite their difficulties, the depth of the pair's association at this time is nowhere so evident as in the epithet they use to address each other – "vous autre" – a term now freighted not only with the respect of the formal "vous," but a coded declaration that each is, for the other, the primary Other, the consciousness which cannot be known, but against which the self must be maintained and measured.

The pretence of something that might be construed as normality ended when the shooting war began. After a chaotic period of twenty days, during which Sartre's unit was moved repeatedly to evade the German Army (and, finally, was deserted by its officers who simply walked off into the woods waving a white flag), Sartre was taken prisoner with a large group of French soldiers.[16] Shut in an attic for several days without sufficient food, Sartre was then taken to Baccarat, where, in the tents that made up the prison camp, and after shaking himself out of a stupor caused by a combination of despair and hunger, he picked up his writing. With his usual disdain for his physical conditions and with his ingrained ability to work almost anywhere, Sartre used his time in the camp to work on the novel he had started during the Phoney War. He also began to put Beauvoir's ideas to work by writing a section of *Being and Nothingness*. Sartre hoped for an early release, a hope that was quashed when he was transferred to Stalag XIID at Trier in August 1940. In the Stalag he was soon working as an interpreter in the infirmary and cultivating his association with the intellectual priests in the camp. Still later, Sartre managed to get himself assigned to the group set aside as the Stalag's artists. In a strange parody of his peacetime existence, Sartre continued his intellectual life. He spent most of his days reading and writing. He tutored one of the priests, Father Marius Perrin, in phenomenology. Sartre even lectured in the Stalag: his talk on the treatment of death in Rilke, Malraux and Heidegger seems to have been a success with his fellow internees. For Christmas 1940, Sartre persuaded the priests to allow him to set up a performance of a play he had written which developed a story tangentially connected to the Nativity, but which was, more importantly, a pretext for a covert dramatization of the possibility of resistance to a conquering power. Sartre, who acted in the play, was elated at its success. It was a bizarre debut – and a telling one – for a playwright who was to figure strongly in the making of the French post-war theatre. Elements of *Bariona*, Sartre's prison-camp play, with its oblique but intense engagement with political issues of the closest personal importance, would be recycled in his later scripts. Sartre, whose correspondence with Beauvoir had been interrupted for only a month and a half in June and July of 1940, wrote confidently to her about his theatrical debut.[17] This dramatic skill was one he intended to cultivate, with all the strength of an ambition that had been with him since his histrionic childhood, when he was free again. Like Beauvoir, Sartre kept himself from despair during his captivity by insisting on his ability to imagine an "afterwards" to his circumstances. He insisted on assuming that he would soon be free.

That freedom came in March 1941. The precise details of Sartre's exit from the Stalag remain obscure, but, late in the month, he appeared in Paris, seemingly out of nowhere, leaving a note at Beauvoir's hotel that she was to meet him at the Café des Trois Mousquetaires. To Beauvoir's anger, she learned that Sartre had already been in Paris for two weeks. He had been sent from the camp with a group of other discharged prisoners who had been kept in a barracks at Drancy. One afternoon in late March, his name had been called. He was put on a train and found himself in the confusion of the Gare de l'Est. After his life as a prisoner, Sartre found his return to the relative freedom of Occupied Paris completely disorienting. He had a difficult time adjusting to his changed circumstances; as he explained later, his time in the prison camp had been like living in

> a sardine can, where I had experienced absolute proximity. My skin was the boundary of my living space. On my first night of freedom . . . in my native city . . . I pushed open the door of a café. Suddenly, I experienced a feeling of fear – or something close to it . . . I was lost; the few drinkers seemed more distant than the stars. I had rejoined bourgeois society, where I would have to learn to live once again "at a respectful distance." This sudden agoraphobia betrayed my vague feeling of regret for the collective life from which I had been forever severed.[18]

It was the physical enactment of individualistic isolation that struck Sartre most on his return to Paris:

> I had the strange impression that people were isolated from one another, whereas in the prisoner-of-war camp we touched each other all the time . . . this sort of contact was, as it were, the superficial impression of a much more profound contact . . . which I first experienced there.[19]

Beauvoir was at first baffled by the change in Sartre. For several days she felt, as she had never done before, that she could not really make contact with him. He seemed to have undergone, more than anything, a *moral* change. Sartre was stern and critical of the compromises she had made to survive in the occupied city. Neither had a clear idea of the other's recent ordeals, and, says Beauvoir, for a time they felt that "the other was speaking in a completely different language." Sartre was shocked that she had signed a statement that she was neither a Freemason nor a Jew in order to keep her teaching post (which, in a grand stroke of irony, was now at Sartre's old school, the Lycée Henri IV). He was equally appalled that she, like everyone else, bought food on the black market when she could get it. At first, Sartre did not understand that simply surviving under conditions of occupation involved some degree of complicity with one's conquerors. And while he was reluctant to talk about his experiences in the Stalag, he was fiercely determined to find some way of acting against the Germans. Beauvoir gave him the news of their friends: Nizan had been killed in the early days of May in the preceding year; Aron had fled to England; Bost had been wounded but was now teaching in Paris; Stépha and Fernando had managed to get to America. The news scarcely touched him; he was "armoured with principles" and his sole thought was of organizing a Resistance group. Beauvoir, always the more practical of the two, was sceptical: she felt that they were too isolated, too inexperienced, and too unimportant to affect the course of events. They had no access to special information and their major contacts were their friends, who were mostly their students and their lovers who needed their help to even survive.[20] Sartre, however, was adamant. If his time in the camps had convinced him that he was part of a collective history rather than a solitary individual, it had also indicated to him, for the first time in his life, the importance of specifically political action. It is in keeping with his characteristic confidence and ambition that his very first chosen sphere of action should be that of the difficult and dangerous organization of an underground group.

Reinstated in his post as a teacher at the Lycée Pasteur, Sartre spent a good deal of his energy between Spring and October of 1941 on his Resistance group, Socialisme et Liberté. He and Beauvoir organized the group around Bost, Merleau-Ponty, and a number of other friends. Theirs was one of several such groups forming in the Paris area with, as Beauvoir put it, "very limited effective strength, and an extraordinary lack of common caution."[21] Accounts of the group's activities indicate both its farcical amateurishness and its sincerity. Bost walked through the streets of Paris openly carrying the group's duplicating machine. Its secret directory was accidentally left in the underground and recovered the next day in the Métro's official lost-and-found office. Nathalie Sorokine simply flung copies of their pamphlets broadcast from her bicycle basket, and, with her characteristic taste for danger, enjoyed handing them to German soldiers whom she guessed could not read French.[22] For the most part, however, the group merely held meetings and argued about the nature of the ideology that was to dominate post-war France, in which the engaged intellectual was to play a significant role. In all this, Beauvoir tended simply to watch, picking out any details that might lead to practical action. Sartre, with his new political awareness now his major inspiration, threw himself into the composition of a constitution for post-war France. There is no doubt about Sartre's passionate dedication to his political aims; the problem was that his efficacy was pathetically limited. In an attempt to bolster the group, he and Beauvoir slipped into the Free Zone in the south of France for a cycling holiday during which they tried, without success, to recruit both Gide and Malraux to their group. With Socialisme et Liberté obviously going nowhere, Sartre became discouraged and turned his attentions back to his writing and to his new teaching post at the Lycée Condorcet, which was to last until 1944. He now understood what Beauvoir could not explain to him in his first flush of outrage when he had escaped from the Stalag. As he wrote in *La France libre* in 1945:

> it must not be forgotten that the Occupation was part of everyday life. When a man was once asked what he had done in the Terror . . . he replied: "I went on living." We could all make the same reply today. For four years we went on living and so did the Germans, in the midst of us, submerged and drowned in the unanimous life of the big city.[23]

In 1941, the other small Resistance groups which were active in Paris wanted nothing to do with Sartre. He was judged too voluble to be trusted.[24] Further, he yielded to the temptation, against Beauvoir's advice, to write on *Moby Dick* for the paper, *Comoedia*, which was published under Nazi censorship, despite the fact that Resistance intellectuals, as a group, refused to supply material to the collaborationist press. When he saw what the paper was, Sartre severed his links with it, but he had made his mistake and it lessened the trust accorded him. From this time onwards, Sartre, who was to become one of the most important figures on the intellectual political scene later in the decade, demonstrated a mixture of intense commitment and practical naiveté. Constitutionally incapable of any role but that of leader, yet hating hierarchy, and incapable, too, of bending for long to the pragmatic expediencies of politics, Sartre was to find his political niche in his writing rather than in other forms of political praxis. In this, though in unexpected ways, his influence was to become incalculable.

In 1941, with Socialisme et Liberté disbanded, Sartre turned to the writing of *Being and Nothingness* and to the composition of his play, *The Flies*, which, like *Bariona* and *She Came to Stay*, would be a fictional work with a coded subtext intended to enter the public domain under the eyes of oppressors.

As anti-Jewish and pro-Nazi propaganda poured out of the Vichy government's headquarters, Beauvoir read the anti-libertarian messages in the light of her personal experience. Bigotry, she said, was something she "knew only too well," it was "the same violent prejudice and stupidity that had darkened my childhood." That message was now writ large; the shadow of oppression "extended over the entire country, an official and repressive blanket." She said that she "was at last prepared to admit that my life was not a story of my own telling, but a compromise between myself and the world at large."[25] Always a realist, Beauvoir decided that railing against injustice was pointless: one either had to find ways around the adversities or to put up with them. In the light of her new apprehension of the relationship between the self and history, she began to develop, in addition to and in harmony with her ontology, an ethics that would concentrate on the individual's responsibilities in a very real world.

Meanwhile, Beauvoir's immediate attention was drawn to her family and her friends. She worried about her sister who was stranded with her husband, Lionel de Roulet (who had been Sartre's student at Le Havre along with Bost), in Portugal for the duration of the war. Georges de Beauvoir died in July 1941. Beauvoir helped him fend off the attention of the priests whom he did not want mixed up with his death any more than with his life. His daughter was "amazed" at "the peaceful way he returned to nothingness."[26] Beauvoir sat with him through "his last moments": she took the death quietly as one more shock amid the general horrors of the time. The manner of his death gave her a good deal of comfort, as did her mother's surprising confidence in making a new life for herself as a widow. The titanic figures of Beauvoir's childhood were now neutralized into pathos.

Despite the grimness of the situation, day-to-day life continued with a complex mixture of pleasure and pain. A new friendship, initiated by Nathalie Sorokine, between Sartre, Beauvoir and the sculptor Giacometti, marked the beginning of an important and lasting relationship for all three. A serious accident on her cycling trip with Sartre left Beauvoir battered, and minus a tooth, but less afraid of death than previously. With their resistance group dissolved, Beauvoir, like Sartre, threw herself into her writing. By now she was working on her second publishable novel, *The Blood of Others*, which, with its emphasis on the tension between general and personal responsibility in an underground resistance group, spoke graphically to the historical situation of the war. As well as writing, Beauvoir and Sartre settled down to the serious and difficult business of keeping alive the circle of dependants they had collected. The couple had always been generous, and from this time onwards they financially supported a large number of associates. During the war, for the only protracted period of her life, Beauvoir took over the immensely difficult and time-consuming task of securing and cooking food for herself, Sartre, Bost, Olga, Wanda, and Nathalie, and, for a time, for Nathalie's lover, a Spanish Jew, Jean-Pierre Bourla, another former student of Sartre, who was subsequently rounded up and killed by the Nazis. The pair concentrated on survival, on waiting for possibilities to improve, and on caring for the younger members of the "family," whose lives were now in their hands. In addition to this intimate circle of younger dependants, Beauvoir was helping to support her mother. She would continue to give her money monthly until Françoise's death in 1964. Although they had no children, Sartre and Beauvoir, from this time forward, financially supported an increasing number of people. As "the couple" at the centre of a pseudo-family often held together by their commitment, despite (as Deirdre Bair puts it) "fractious intrigues, unsatisfied appetites and economic insecurities," Sartre and Beauvoir gained as much as or more than they put into this idiosyncratic arrangement. As Beauvoir herself said, their "cult of youth" kept them young. Their younger associates provided them with a stimulus they could not otherwise have secured.

> Certainly we fed on [their] vitality. It nourished us. We preferred it to the talk of our contemporaries, with their endless descriptions of house furnishings and the intelligence of their babies and their smugness in the bureaucratic security of their work.[27]

The youth of most of the family enhanced their sense of freedom: it also gave them power over those they recruited to their own rebellious way of looking at life. As teachers, guides, and lovers, Sartre and Beauvoir were always to align themselves with youth, possibility, critique, and change. If they sometimes looked foolish doing so, it was a price they were more than willing to pay.

The winter of 1942/3 was cold, long, and difficult. Rationing tightened, and food continued to be the overwhelming preoccupation of all Parisians, Sartre and Beauvoir included. Simply to stay alive, they spent so much time in the Café Flore, grateful for its heat, light, and companionship, that Bourla remarked, "When they die . . . you'll have to dig them a grave under the floor."[28] Their living conditions were troublesome for a variety of reasons. In the autumn of 1942, Beauvoir and Sartre lost their rooms at the Hotel Mistral, in which they had lived on separate floors since Sartre's release from the Stalag. Beauvoir piled their things into a handcart and hauled them through the streets to the Hotel Aubusson, near the Pont Neuf. Beauvoir had never been afraid of austerity – indeed her and Sartre's lives, except for their passion for travelling and their habit of eating in restaurants, were always frugal in terms of material possessions – but the Hotel Aubusson was sordid even by her low standards. It was "a filthy dump, with an icy stone staircase that reeked of damp and innumerable other odours," and the other residents of the hotel included a prostitute who battered her son and a tenant who collected excrement in the cupboards.[29] Beauvoir was to live there for two years. Almost immediately after the couple had settled, Wanda moved herself in with Sartre, while other members of the family took rooms on the floor below. Beauvoir, exiled from Sartre's room by the younger Kosakiewicz sister (Wanda made scenes whenever Beauvoir entered), was left, she said, "mostly celibate" at this time. Her job was to take care of the family's needs. As she came to terms with both her jealousy and the housewifely burden placed on her for the first and only time in her life, she comforted herself with the "satisfaction" and "gratification" she received from others in non-sexual ways. Looking back on this period she remarked, "Sexual completion is not everything. There are other ways to care for people, to love and be loved."[30] Nevertheless, the situation was difficult for Beauvoir; only the enforced solidarity of the war seems to have kept the potentially inflammable situation under control.

For all its difficulties and deprivations, 1943 was a crucial year for both writers. In January, Sartre was invited by a communist friend from the prison camp to join the Comité National des Ecrivains (CNE), an important group for Resistance intellectuals. He also contributed to *Les Lettres françaises*: this clandestine paper, whose circulation reached 12,000 by the autumn of 1943, covered both arts and politics, and was eagerly passed from hand to hand along with *Libération*, *Combat*, and *L'Humanité* as an underground source of news that had not been processed by the collaborationist Vichy press. The paper tried to build morale, shame writers who worked for the censored French media, and endorse the work of Resistance artists. Beauvoir was not included in the CNE as she had not yet published a book (and besides, she felt that the undertaking was too bureaucratic for her taste, unlike Socialisme et Liberté, which, for all its faults, at least had the virtues, from her point of view, of being "improvised and hazardous"), but she watched Sartre's activities with interest and a little regret at her exclusion.[31] Most importantly, said Beauvoir, she was "very glad we had emerged from our isolation, all the more so since I had often felt how tedious Sartre found a life of passive inaction."[32] They were finally linked to the growing Resistance in a minor, but recognizable way. Sartre, in particular, had moved into the world of successful political action.

The year 1943 was, in fact, one of the most significant years for Sartre's reputation as a writer. *Being and Nothingness*, with its manuscript carefully worked over by both Sartre and Beauvoir, appeared in June – dedicated, in what must be the most understated gesture of thanks in the history of philosophy, to "le Castor." Its publication coincided with their friend Charles Dullin's production of Sartre's first professional play, *The Flies*, which featured Olga in the most important role of Electra. Sartre had turned to the composition of the play in 1941; with the break-up of

Socialisme et Liberté, said Beauvoir, "he obstinately settled down to the play he had begun, which represented the one form of resistance work still open to him."[33] Building on the strategies he had initially used in *Bariona*, *The Flies* is a dramatization of the enactment of liberty under crippling circumstances. Despite the slight camouflage afforded by the classical subject-matter of the plot – a treatment of Orestes' revenge on Aegisthus and Clytemnestra for the murder of Agamemnon, in which Orestes not only performs the requisite act of revenge against parricide, but also serves as the scapegoat for the plague of flies that has tormented the citizens of Argos since the slaughter of their rightful king – the play was immediately read as a thinly veiled allegory of the situation in Occupied Paris. Michel Leiris praised the play in *Les Lettres françaises* for its political impact. Merleau-Ponty hailed it as "a drama of liberty."[34] In *The Flies*, as in *Bariona*, Sartre not only had the pleasure of making a public political gesture under the noses of the Germans, but he also learned a great deal about the practicalities of stagecraft by watching the rehearsals and taking in Dullin's directorial methods.[35]

On its first appearance, *The Flies* was not a substantial commercial success. Its initial run lasted for only twenty-five performances. But it brought Sartre into the public eye (Beauvoir was delighted to report to Sartre a conversation about his play by strangers that she had overheard on a train).[36] *Being and Nothingness*, too, took some years to find its readership.[37] In spite of the lack of immediately spectacular success, Sartre's career as a writer was now established. His connections with other writers grew quickly. Camus introduced himself to Sartre at the dress rehearsal for *The Flies*. This was the beginning of another significant friendship of the 1940s for both Sartre and Beauvoir. In the autumn of 1943, Sartre wrote his next play, *No Exit*, in two weeks, and persuaded Camus both to direct the play and to take the leading role. Sartre's activity at the time was prodigious. By the beginning of 1944, he was a member of the Comité National du Théâtre as well as the CNE.[38] He finished his novel, *The Reprieve*, by the end of 1943, as well as contributing regularly to the underground Resistance papers. He was not only satisfying his multifaceted ambitions as a writer, but he was also acting with the political engagement which he now believed was the writer's primary duty. Further, history seemed to be turning: the war was going badly for the Germans. Not only the present but the future seemed full of hope.

If the publication of *Being and Nothingness* was, in time, to secure Sartre's reputation as one of the great voices of twentieth-century philosophy (and one must remember that Beauvoir not only allowed this to happen, but assisted Sartre in bringing it about), and *The Flies* marked his debut as one of the key playwrights in post-war European drama, Beauvoir presented her experience of 1943 as a combination of disaster and triumph. In the spring of the year, Nathalie Sorokine's mother insisted that Beauvoir use her influence with her daughter to break up the relationship with the Jewish Bourla. With her respect for individual freedom and her hatred of anti-Semitism, Beauvoir refused. Madame Sorokine took revenge by reporting her to the educational authorities with the charge of corrupting a minor. Beauvoir was struck off the register of the university teaching roll and thus lost her post, and, with it, her source of income. Sartre, by now brushing aside the high idealism of the period immediately following his release from imprisonment, found her a dubious job through a collaborationist friend. It involved working with Bost's older brother, Pierre, making radio programs about traditional French festivals for the hated Radiodiffusion Nationale. The couple's pragmatic need to survive overtook their scruples. Beauvoir's pay was excellent and contributed a good deal to the voracious family's coffers. Further, Beauvoir enjoyed collecting material for the broadcasts, despite her uneasiness about her employment in the German-controlled media. Her summer, she said, started anxiously as Wanda claimed for herself three weeks of the holiday Beauvoir was to have spent with Sartre, but the great triumph of her year came with the publication of *She Came to Stay* during her vacation. Finally a published author, after so many years of trying to find her way into print, Beauvoir's first novel was a commercial success, and the good reviews it received excited her. Gallimard printed 23,000 copies and she was mentioned – like Sartre on the publication of *Nausea* – as a candidate for the Prix Goncourt.[39] Beauvoir needed public validation more than she thought. More than anything, it was the knowledge that her work had been *read* seriously that pleased her. Toward the end of August, Sartre went to Paris for a meeting of the CNE. When Beauvoir met him at the station at Angers, he hurried towards her waving a paper. The first review of *She Came to Stay* had just appeared in the dubious *Comoedia*. As Beauvoir recalled, her reaction was ecstatic: "No article," she said, "had ever pleased me so much." She felt that "at last" she "was fulfilling the promises I had made to myself when I was fifteen." The sight of the first review of the novel was particularly important to her, as she said:

> It is not very often that one unequivocally achieves a long-cherished ambition. Here was a review, printed in a real paper, to assure me in black and white that I had written a real book – that I had become, overnight, a real writer. I could not contain my joy.[40]

That this validation first occurred in the very journal that had undermined Sartre's credibility with the Resistance seems not to have even crossed Beauvoir's mind. In a rush, she found her public. The novel was widely, and favorably, reviewed: Beauvoir received letters about it from Gabriel Marcel, from Cocteau, from Mauriac. Everyone seems to have read it as a racy, and highly personal, commentary on the Parisian scene; no one apprehended its philosophical depths. But the book served its purpose of launching Beauvoir as a public figure in one stroke. When she returned to Paris with Sartre at the end of a cycling holiday that concluded with a stay at Madame Morel's, she was received as a literary celebrity.

Sartre and Beauvoir were now both rising stars on the Paris wartime intellectual scene. One of the benefits of their new position was that it allowed them to make new contacts, of which they were badly in need. The war had scattered their friends: Beauvoir noted that they now saw almost no one outside the enclosed, and rather too incestuous (and demanding) family.[41] They felt fortunate as their world opened up. Through their new, lifelong friends, Michel and Zette Leiris, they met Picasso, Lacan and Georges Bataille. Raymond Queneau and Mauriac introduced themselves. Sartre and Beauvoir began to be "in demand" as a public couple. As Deirdre Bair observes, they had gone from being just "a couple" to a "writing couple" and were now headed toward becoming a "professional couple."[42] The adjustments and temptations this new – and shared – role brought to them reached beyond anything Beauvoir had imagined after the publication of *Nausea* in 1938, and the way they addressed the long-term pressure of this situation was different for each.

However, in 1943, both were immersed in their projects. After they moved out of the hated Hotel Aubusson to new and more pleasing rooms in the Hotel Louisiane on the rue de Seine, Beauvoir honored the lessons of the war by constructing an ethics that would address the need for action in specific material circumstances. In her philosophical essay "Pyrrhus and Cinéas," as in *The Blood of Others*, which she completed that year before beginning her third novel, *All Men are Mortal*, Beauvoir was working through what she called her "moral period."[43] During this time she codified her current similarities and differences with the ideas she had held at the time of the composition of *She Came to Stay* and *Being and Nothingness*. More importantly, she worked out an ethics with a material base. This seemed to her a new and pressing task, given her experience of the war, and her now deep sense of the impact of history on the individual. In "Pyrrhus and Cinéas" she clearly defines her moral starting point:

> This then is my position in regard to other men; men are free, and I am flung into the world amid these alien liberties. I have need of others for once I have passed my own goal my actions would recoil on themselves inert, useless, were they not carried by new projects toward a new future.[44]

This opening out to the world, to the need to reach beyond individual projects to the generality of humankind, and to the need for reciprocity, was to provide Beauvoir with the grounding for the moral principles on which she would base both *The Second Sex* and *Old Age*. At the time of their composition it looked as if "Pyrrhus and Cinéas," and Beauvoir's later book on morality, *The Ethics of Ambiguity* (1947), provided the missing companion volume of ethics to the ontology of *Being and Nothingness*. Where Sartre and Beauvoir's earlier work had analyzed individual consciousness and its construction, Beauvoir now passionately wanted to "provide existentialist morals with a material content."[45] In her conversations with Sartre, Beauvoir upheld against him her idea of the importance of establishing "an order of precedence" among various possible ethical situations. Her concern was to not leave freedom of ethical choice in a vacuum, but to measure it against individuals' capacity to take whatever positions of privilege they are afforded and use them to construct "a new future," not only for the self but for each other. Her main point was that an "activity is good when you aim to conquer those positions of freedom, both for yourself and for others: to set freedom free."

Beauvoir's line of thought was particularly engaged with her and her friends' historical moment. The chief intention of the new intellectual grouping that was forming around Sartre and Beauvoir in the last years of the war – Camus, the Leirises, Raymond Queneau and his wife, Merleau-Ponty – was a source of both strength and mutual resolve. This intention was nothing less than to be the source of a new and ethically informed ideology that was to shape post-war French culture. At first, as Beauvoir said, the group fed off its unity:

> We agreed to remain leagued together in perpetuity against the system and men and ideas that we condemned. But their defeat was imminent, and our task would be to shape the future that would then unfold before us: perhaps by political action, and in any case on the intellectual plane. We were to provide the post-war era with its ideology.

The group was to succeed in this ambitious aim better than it had any right to expect. In 1943, the major point of personal importance for Beauvoir was that she was now a writer with an identity in the public arena, and that, with the course of the war turning, it looked as if "a day would come when the future was open again."[46] She felt her possibilities for happiness, which she thought gone forever, come flooding back.

Indeed, all Paris seemed to feel the same as 1944 brought an increasing certainty that liberation was inevitable. Beauvoir was again disappointed when Wanda claimed Sartre for the Christmas holidays he had promised to spend with her: Beauvoir covered her annoyance and went skiing with Bost. The couple were drifting still further apart in their physical intimacy. Sartre's random womanizing grew more obsessive, and his behavior merged with the rising saturnalia that took over the couple's widening circle of friends in Paris. The year was one of "fiestas" – parties that featured hard drinking and dancing till dawn in defiance of the curfew. They were a sign of the euphoria that accompanied the general feeling that the war was in its last throes.

This was a crucial moment for Beauvoir and Sartre's association. Now, Beauvoir worried less about the distraction from her that other women posed for Sartre, than about the depth of his attachment to Camus. For some years, the couple's "contingent" lovers had been physically more important to them than their "primary" union. But (and despite Beauvoir's attachment to Bost) their intellectual bonding had been stronger than ever (and Sartre's debts to Beauvoir never greater) during the composition of *Being and Nothingness*. Camus seemed to challenge that bond. For the first time since his years at the ENS, a convincing intellectual rival for Sartre's attention had emerged. Further, unlike Sartre, Camus had no time at all for ideas of sexual equality. "'Camus couldn't stand intellectual women,'" said Beauvoir in 1982. "'His usual tone of voice for me was, to say it politely, ironic mockery. That stops short of what it really was – generally insulting.'" To her annoyance, Beauvoir felt she was in competition for her life-partner: "'We were like two dogs circling a bone. The bone was Sartre, and we both wanted it. I cared about it more than Camus. After a while he got bored and went sniffing after other things.'"[47] Beauvoir hid her anger over Camus while Sartre was alive. After his death, and even after so many years, her irritation still showed. However, as was usually the case, Beauvoir's will directed the course of her and Sartre's relationship, which held together despite the threat that Camus posed.

What Camus brought Sartre was his energy. The two men fed each other's taste for grand plans in politics, writing, theatre, and journalism. They shared the same rather childish sense of fun that, whatever Beauvoir thought in general, helped provide amusement for the couple and their entourage at a time when it was badly needed. Camus, said Sartre, was "'probably the last good friend I had.'"[48] Plans for Camus, Sartre, Beauvoir, and Merleau-Ponty to work together to produce the Gallimard *Encyclopedia*'s new volume on ethics, envisioned as their "joint manifesto" for the post-war future, came to nothing, but the existence of the plan demonstrates just how close Sartre and even Beauvoir felt they were, intellectually, to Camus toward the end of the war.[49] They later found that they were mistaken in this, but, in spite of Beauvoir's retrospective hostility, this new association provided both of them with another associate against whom to test their ideas, and to play roles other than those which pertained in the family.

Sartre's other important friendship of the period was with Jean Genet, who, like Camus, wrote for Gallimard. Beauvoir was impressed with his work, and, for Sartre, Genet's writing presented him with what he thought of as his last new literary enthusiasm of his life.[50] If, in Camus – with his Alsacian father, who had died before he was a year old, and his impoverished Spanish mother, who had done what she could to raise her son in hard circumstances – Sartre could both find a partial reflection of himself and discover, at close quarters, his first working-class hero in the handsome Algerian, then in Genet, the attractions were even more anarchically prominent for the philosopher who defined himself (with better cause than anyone in his lifetime perhaps understood) as both a rebel and a thief. In Genet – the criminal turned writer – Sartre believed he saw a mirror of himself.[51]

Cocteau had "discovered" Genet in prison, and Sartre was flattered when Genet introduced him to the legendary figure who had been interested by *No Exit* when it opened in June 1944 (without Camus, who felt, on reflection, that he was not experienced enough to take on a serious Parisian production of a play).[52] Sartre was now approaching 40 and Genet's toughness appealed to him as a signal of potency and youth. His links with the theatre were also important. The theatre had become one of the focal points for the couple's activities. Sartre lectured on drama during this period, and Beauvoir, who had been entranced with the theatre since the production of *The Flies*, finished her only (rather unsuccessful) play, *The Useless Mouths* – another work on the need for individual and collective responsibility, which focused, presciently, on the place of women, children, and the aged in the body politic – in the summer of 1944. Sartre and Beauvoir were strongly linked to the Parisian stage, not only through Olga and Wanda, who were now both actresses, but through Simone Jollivet and her partner, Dullin, who had given them the entrée to the theatrical circles which formed part of their expanding social and professional networks.

Before Paris was liberated, in August 1944, Sartre and Beauvoir were warned to move out of their lodgings after a member of the *Combat* Resistance group, with which Camus was deeply involved, had been arrested. They saw their first, but not their last, experience of being displaced from their rooms for political reasons, as a game, a game that merged into the delirium of joy that swept the country – despite the viciousness of the last phase of the war – as France was set free.

Camus commissioned Sartre to write a series of articles on the liberation of Paris for *Combat*, a joyous task as the Resistance papers now emerged from underground to celebrate the rout of Nazism from France. Beauvoir actually wrote the articles, which appeared under Sartre's name – Camus's reaction to a woman doing war reporting for *Combat* can only be imagined – between her searches for scarce provisions. Many years later, she explained that she thought this was "'a perfectly reasonable thing to do . . . Anyway, what did it matter whose name it was? Someone had to write them.'" When asked if she had, on other occasions, written things that were published under Sartre's name, she evaded the question: "'Maybe sometimes. I forgot.'"[53] Whose name featured on the by-line of the *Combat* articles certainly mattered very little at the time to the couple, who threw themselves into the general jubilation of the ending of the war in Paris. They watched de Gaulle's victory march down the Champs Elysées – Sartre from a balcony in the Louvre; Beauvoir, with Olga and the Leirises, from the Arc de Triomphe – and cheered and laughed with the rest of the city, mad with joy in the "magnificent, but chaotic, popular carnival show" of victory.[54] But their eyes were steadily on the future. As Beauvoir explained, during the war,

> We were looking ahead to the peace, no matter who the victor. We were planning for our future in France after the war. When we decided what that should be, then we would begin to work to assure it . . . intellectuals had the obligation to think of these things . . . Anyone who wasn't there does not have the privilege of criticizing those who were.[55]

7

EXISTENTIALIST HEROES

"'What about you, madame? . . . Are *you* an existentialist?'" In 1943, Beauvoir was startled to be asked this question on being introduced to Jean Grenier, by Sartre, in the Café Flore. She had no idea what was meant. Beauvoir was familiar with the term "existential philosophy," which was associated with Kierkegaard and Heidegger, but, she said, she "didn't understand the meaning of the word 'existentialist', which Gabriel Marcel had recently coined."[1] This surprising question was the first sign of the storm of notoriety that broke over Sartre and Beauvoir in the aftermath of the war, as the two writers found themselves transformed from struggling intellectuals into focal points of attention from the international media, and superficial role models for the first post-war youth cult.

For a brief period after the Liberation, it seemed to the intellectuals of the Resistance that an entirely new social order, put together under their auspices, could replace the pre-war social formations in France. The hope, said Beauvoir, during the "orgy of brotherhood" that followed the Liberation was aptly summed up in *Combat*'s motto, "'From Resistance to Revolution'."[2] That hope failed to take into account the political ambitions of de Gaulle, who bypassed members of the Parisian Resistance when appointing his first post-war administration. The General favoured industrialists, pre-war politicians, and, unsurprisingly, his own political associates over the leaders of the Resistance, and he worked quickly to consolidate a strong state which would support him in a position of maximum control. The Resistance intellectuals could match him neither in political skill nor widespread support.[3] They also reckoned without the polarization of left and right which accompanied the immediate rise of the Cold War following the Second World War. But, for a time, in the euphoria of liberation, and with the clear-cut moral and political commitments of the Resistance still dominating the minds of French intellectuals, it looked not only as if a new order could be created, but also as if it would be they who created it. Sartre and Beauvoir, with their strong ties to the Resistance press, were at the heart of this storm of optimism which lasted for only a few months, but which was to provide the impetus for their writing for many years to come.

Sartre now saw his chief task as communicating as directly as he could with as large a public as possible. On the day after the Liberation he gave an interview to *Carrefour* in which he outlined his intention to write for the theatre and the cinema in his pursuit of a mass audience. "This year," he said, "I'll write a play on a contemporary subject . . . Through present events we'll try to bring out the unchanging relations between men. The eternal ought to struggle free of the everyday."[4] Excitedly, Sartre saw the chance to bring together his philosophical, literary, and political interests in ways that would have far-ranging effects. He was moving quickly toward his idea of *littérature engagée*, the hallmark of his post-war writing. Literary works had to speak to the generality of humankind in ways that illuminated the historical situation of their composition and that illustrated the inevitability of freedom to individuals who might otherwise feel trapped or powerless within their circumstances. As always, Sartre's project was ambitious, and he threw himself into his dramatic writing with characteristic energy. The results were highly successful. As Michel Contat and Michel Rybalka note, Sartre's plays dominated the French stage from the end of the war until 1951, and his international reputation was, at first, overwhelmingly grounded in his theatre work. Sartre's subsequent fame was probably due "far more to his plays than to his novels, essays or works on philosophy." The plays were accessible, and Sartre put a great deal of store in their publication in book form as well as in their performance.[5] They remain the most widely read of all his work and Sartre seems to have understood their potential popularity very well. It was what he was aiming for. He wanted a mass audience, and with his plays he got it. In a lecture given on June 10, 1944, just two months before the Liberation and a few days after the dress rehearsal for *No Exit*, Sartre outlined the principles that were to guide him as a playwright. The audience for the lecture included Camus (who was to give the next lecture in the series), Cocteau, and Salacrou, all of whom joined the discussion that followed Sartre's talk. It was an important occasion, one that helped to set the direction of the French post-war theater. Sartre began by distinguishing the difference between the cinema, the novel, and the theater as narrative art forms. Unlike the novel, the theater must assume a distance between the audience and the characters' perceptions. Where the novel involves the reader's "complicity" with the author's wish to create characters whose perceptions become those of the reader, the theater does no such thing. Likewise, the cinema – in which the camera determines the content and angle of vision – breaks down the distance between the audience and the film-maker. Cinema directs the gaze of the audience in absolute ways. Therefore, where naturalism can work in the cinema, and psychological interest can be made the mainspring of the novel, theater must proceed by other means. It works through gesture and action, whether that action is verbal or physical. It has no other tools. This definition of the theater's nature moves Sartre into a position from which he can explain his idea of the purpose of the theater. The "act," he says, "is *ipso facto* devoid of psychology," rather, it is "a free enterprise." It is radically unexplained and open to interpretation by the audience, which is not given, and cannot be given, a transparent guide to an action's meaning. This emphasis on freedom in the theater leads Sartre to his account of the theater's deep purposes which are moral, in the sense that theatrical acts are connected to debates about the nature of rights, and the possibilities of interpreting the varying claims of those who invoke them. "The true ground of theater," argues Sartre, is not psychological, but is constituted in "watching a conflict of rights." The theater, he repeats, is "a sort of ring in which people battle for their rights." The strength of Sartre's passion in believing in the drama as a location for moral debate which is ultimately political, is evident in his comments on the place of language in the theater:

> the audience is not in the slightest interested in what goes on inside a character's head, but wants to judge him by everything he does. It is not concerned with some sort of slack naturalistic psychology; it does not want speech used to depict a state of mind but to commit. Speech in the theater should express a vow or a commitment or a refusal or a moral judgment or a defense of one's rights or a challenge to the rights of others, and so be eloquence or a means of carrying out a venture, by a threat, for instance, or a lie or something of the sort; but in no circumstances should it depart from this magic, primitive and sacred role.[6]

Though he would later argue, with some disgust, that the bourgeois control of the theater was almost absolute, Sartre hoped, in the 1940s, that his plays would transcend class barriers.[7] He was convinced that one could address the masses through the drama in ways that spoke powerfully to "the most general preoccupations, dispelling . . . anxieties in the form of myths which anyone can understand and feel deeply."[8] Sartre intended to do just that, and he continued to work to create a theater that would teach the lessons of freedom to his fellow countrymen and women who had so recently felt themselves enslaved.

*   *   *

The theater was only one of the areas to which Sartre and Beauvoir now excitedly turned their attention. In the autumn of 1944 they abandoned their teaching posts. On the strength of his royalties from *No Exit* and his expected earnings from writing for the cinema, Sartre gave up his teaching post at the Lycée Condorcet at the beginning of the autumn term in 1944. He applied for a leave of absence from the university's national teaching roster to devote himself exclusively to his writing. Beauvoir, whose name had been cleared and reinstated on the national teaching roll after the Liberation, did the same. Sartre was now earning enough money to support them both, and though her financial dependence gave Beauvoir a few uneasy moments, in general, she relished the chance to give her entire attention to her writing. She felt that writing gave her a "moral autonomy" she could not secure as a teacher.[9] Besides, since she had begun *She Came to Stay* in the late 1930s, she felt, with good reason, that she had come into her own as a writer. She had served her apprenticeship, now she wanted to practice her trade. In *Old Age* in 1970, she recalled her early self-dedication to the writer's life: "I remember when I was eighteen how earnestly I wrote in my diary, 'I shall say everything. I mean to tell all . . .', whereas it so happened that I had nothing whatever to say."[10] From the late 1930s on, however, material would never be a problem for Beauvoir as a writer; she found that now she "always had 'something to say'." By 1944, her books, for Beauvoir, "were a real fulfilment."[11] It was Sartre's money that initially made the pair's new lives possible, but then, Beauvoir's ideas had provided a crucial base for his success. Whatever the nature of their joint contribution to their new position, the couple were united in their pleasure in having the confidence to launch themselves jointly into the writers' lives they had always intended to live.

The delight occasioned by the Liberation and by the companions' decision to interrupt their careers as teachers was undermined by the revelation of the full horrors of Nazi atrocities which followed in the weeks immediately after the end of the occupation of Paris. The war was not over, and would not end in Europe until May 5, 1945. Parisians learned of the fire-bombing of Dresden; in November, the V-2s fell on London; there were rumors of new secret weapons in German hands; Beauvoir feared the reoccupation of Paris. Guilt and fear lay under the exuberant surface of the French capital. As Beauvoir noted, the revelation of the full details of the massacres and executions of the recent past laced the joy of freedom with guilt and desolation: "One's new delight in life gave way to shame at having survived." Both the shame and the horror were to deepen later with the opening up of the German concentration camps. Bost, working as a war correspondent for Camus, was one of the first to go into Dachau in 1945. Beauvoir said that he "could find no words to describe what he had seen." The bombing of Hiroshima and Nagasaki in June 1945 was still to come. Despite what they had lived through, Beauvoir and Sartre discovered that they "had known nothing" of the depths of depravity and cruelty of the war. But this darkening of vision was still to come: in the autumn of 1944 the couple and their friends, said Beauvoir, "all sang in chorus our hymn of the future."[12]

The excitement and hopefulness of this time, and the way it fed into Sartre and Beauvoir's personal careers, cannot be overstated. The couple felt themselves at the absolute centre of the reawakening of Parisian cultural life. Opening *Combat* "in the morning," said Beauvoir, "was almost like opening our mail." With his new-found dedication to the idea of speaking to the masses, Sartre was himself determined to found a journal which would speak to the truth of the era in a new way. He and his journalists, wrote Sartre, would "be hunters of meaning, we would tell the truth about the world and about our loves." The first editorial committee of Sartre and Beauvoir's journal, *Les Temps modernes* (the name, said Beauvoir, "was dull, but the reference to the Chaplin film pleased us"), was set up in September 1944.[13] It included Raymond Aron, who had returned from his wartime exile in London; Merleau-Ponty; Camus's associate, Albert Olliver; and Jean Paulan, who had helped to found *Les Lettres françaises*.[14] Camus and Malraux were also invited to join the undertaking: disappointingly, both refused. By now, Sartre was even more intent on his experiment in mass communication. "I conceived," he was to say in 1972, "of a 'total public', something earlier writers had never been able to do." During the war he had come to the conclusion "that there were no purely literary writings," that "all writing is political."[15] The new journal would be the means by which Sartre, Beauvoir, and their journalists would influence the ideological formations of post-war France by commenting on anything that seemed important in their quest for truth. The idealism and openness of the journal soon attracted an interested readership. *Les Temps modernes* quickly became an important organ of the non-aligned left. It engaged Sartre's, and particularly Beauvoir's, attention for decades. The journal provided the couple with a ready-made platform for their ideas and a guaranteed source of publication for their writing. It was successful enough and popular enough to survive its founders. In 1944, the production of the first issue was problematic, for the very practical reason of a paper shortage. Beauvoir dealt with the Ministry of Information, trying to secure the necessary allocation of supplies. Her success, and, indeed, that of the publication in general, delighted her. The project appealed to her on grounds other than those solely connected with journalism and the new security offered by an outlet for her writing. For Beauvoir, the *social* aspects of the journal were a source of pleasure: "already I was beginning to enjoy myself enormously," she said of the inauguration of the journal, "this community of enterprise seemed to me the highest form of friendship."[16] Sartre and Beauvoir's intimate collaboration of nearly a decade and a half was now widening out to include a range of associates who wanted to affect society at large. Sartre and Beauvoir's publisher, the prestigious Gallimard, put up the money for *Les Temps modernes* which did, indeed, take its place as one of the most influential papers on the post-war scene.

That scene opened out with astonishing speed, and when it did, Sartre and Beauvoir were ready. In the latter part of 1944, Sartre was writing *Baudelaire* and *Anti-Semite and Jew*. He sent the two completed novels of his projected tetralogy, *The Age of Reason* and *The Reprieve*, to Gallimard. Beauvoir looked for a director and producer for her first and only play, *The Useless Mouths*. Michel Vitold, who was to become her good friend and casual lover in 1945, agreed to take it on. At the time, Beauvoir entertained serious dramatic ambitions for herself. Elated by Sartre's success in the theatre, and enthralled with her new, first-hand, behind-the-scenes glimpses of stage life, Beauvoir was determined to match Sartre in this new field. It was one of the few projects she undertook in maturity that was to fail.

But failure of any kind was the last thing on the couple's minds as 1944 turned into 1945. Towards the end of November in 1944, Camus arranged for Sartre to be one of eight French journalists to go to America. Sartre was to write reports for *Combat*, and also picked up some work for *Le Figaro*. He was inordinately pleased. And, though Beauvoir was disappointed that she was not going too, she said she also rejoiced, "not only for Sartre's sake, but also for my own, because I knew that one day I was sure to follow him down this new road." For the couple, said Beauvoir, America, with its "jazz, cinema and literature" which had "nourished" them in their youth, was "a great myth." They saw the United States as the country which had delivered France from war, as a place of utopian promise: "it was the future on the march; it was abundance, and infinite horizons," a "magic lantern of legendary images."[17] At least as Beauvoir tells it, the couple seems to have had a severe case of The American Dream. Their unrealistic, even adolescent, enthusiasm for America was, not unexpectedly, to be tempered by their experience of the United States. But when Sartre left for his first-ever flight, in a military plane for New York on January 12, 1945, there was only one shadow over the trip. That drawback was to become more important than might at first appear. There was no postal service at the time between America and France for civilians. For the first time since Sartre's incarceration as a prisoner of war (which was itself the only lapse in their contact since their first compact as lovers), the couple would be almost completely cut off from each other. The only way Beauvoir could get news of Sartre was by reading his articles.[18] Their emerging public roles were generating impediments to their complicated intimacy that they had not imagined. This trip, despite Beauvoir's later statements of joy on his behalf, was Sartre's and Sartre's alone. It meant so much to him that, when his stepfather died on January 21, his mother did not even attempt to get word to him, for fear that the news would spoil his obvious, and single-minded, pleasure.

Beauvoir and Sartre both present Sartre's time in America from January to May 1945 as a watershed for them in a number of significant ways. Sartre's confidence in himself at the time cannot be overstated. He felt that his ambition to earn immortality through his writing had already been vindicated. As he put it toward the end of his life (and it is interesting that in this survey of his achievements Sartre makes no mention of *Being and Nothingness*):

> Until after *Nausea* I had only dreamed of genius, but after the war, in 1945, I'd proved myself – there was *No Exit* and there was *Nausea*. In 1944, when the Allies left Paris, I possessed genius and I set off for America as a writer of genius who was going for a tour in another country. At that point I was immortal and I was assured of my immortality. And that meant I no longer had to think about it.[19]

At the time, however, the journey did not seem quite so grandiose nor Sartre's position quite so assured. Sartre's first American trip was a propaganda mission. It involved a mixture of deeply tedious official tours of cities, factories, dams, and bridges, organized by the Office of War Information, which also arranged for its party of tame journalists to meet President Roosevelt, who died only five weeks later. Sartre was baffled by American cities, unimpressed by American education, and appalled by American conformity. He stirred up trouble almost at once by writing an article which accused the American State Department of buying off French partisan interests in the States. His forthrightness almost got him sent back to France before a month of the tour had passed.

In personal terms, the trip was profoundly important. In New York, Sartre made contact with Stépha and Fernando Gerassi, and was taken to see the most distinguished French refugees, from Claude Lévi-Strauss to André Breton. In Hollywood, he saw Nizan's widow, Henriette, who told him about the campaign of vilification of Nizan after his death by his former comrades in the Communist Party. Sartre was to respond forcefully to these accusations through *Les Temps modernes* in 1947, when the journal would publish a defense of Nizan that marked Sartre's open split with the communist faction in France.[20] But the most significant person Sartre met in America was a woman who worked for the Office of War Information as a broadcaster of French-language programs during the war. Dolorès Vanetti Ehrenreich had been an actress in Montparnasse between the wars. Partly estranged from her husband, an American doctor, she was a former lover of André Breton. Dolorès was, by all accounts, intelligent, charming, talented, and could serve as Sartre's translator and guide in a country where the language was impenetrable to him and the landscape confusing. Almost entirely cut off from Beauvoir, except for brief notes he sent through Camus, and relishing his new persona of foreign dignitary, Sartre became passionately attached to Dolorès, who reciprocated his feelings enthusiastically, without understanding the complex nature of Sartre's connection to Beauvoir. In a conversation in 1974, recalling this period, Beauvoir reminded Sartre of how "immensely attached" to Dolorès he had been. "Furthermore," she said, Dolorès "was the only woman who frightened me. She frightened me because she was hostile." Sartre's other lovers accepted Beauvoir's special place in his life and worked around it. Dolorès did not: she wanted Sartre for herself, and brought him gifts that he valued. Dolorès, said Sartre, "did after all give me America. She gave me a great deal. The roads I travelled in America made a web around her."[21] Beauvoir and Dolorès's competition for Sartre was conducted out of the public eye and on rather different terms than Sartre and Bost's competition for Beauvoir. The two women scarcely spoke of each other to any but intimates until both were elderly. But Dolorès's addition to the Sartre/Beauvoir/Bost equation partially upset the couple's balance for some years. From 1945 until the early 1950s, when the situation was complicated by Beauvoir's own additional serious affair with Nelson Algren, and then clarified by her and Sartre's joint choice of each other over their other lovers, Sartre was, in Dolorès's view, "troubled, unsettled, undecided." The same was, however, true for Beauvoir and Dolorès as well. All three were, said Dolorès, "all these, and much worse."[22] It was a classic case of deep fracture in a long-established relationship, and for a time, it was difficult to know if Sartre and Beauvoir would come through it united on any terms at all.

By May 1945, when Sartre returned from America, Beauvoir had completed a trip of her own as a foreign correspondent for *Combat*. In Sartre's absence, she had been close to Camus, and was pleased when he asked her to go to Iberia to report on conditions in Madrid and Lisbon, where she could visit her sister and Lionel, who was working at the French Institute in the Portuguese capital. Beauvoir's hard-hitting articles for *Combat* and *Volonté*, condemning the social conditions of the poor and denouncing the dictatorship of Salazar, were signed with a pseudonym, Daniel Secretan, so as to avoid embarrassment to her brother-in-law. Hélène, shocked at her sister's appearance after the deprivations of Paris, bought Beauvoir clothes and fed her extravagantly. For her part, Beauvoir rejoiced in being pampered for a change, and returned to Paris triumphant with her bags full of luxuries that had been unavailable in France for half a decade. Her pleasure was undermined when Camus brought the news that Sartre had left the official journalists' tour in America to return to New York to live with Dolorès. He would return late to Paris. Beauvoir, who had felt old at the age of 36, felt betrayed, older still, and unloved. She comforted herself for this serious defection by her partner of fifteen years with a brief affair with Vitold. She was also seeing Maheu again.[23] If her primary relationship looked as if it might be in deep trouble, her contingent love affairs were proceeding with their usual complicated flair. Nevertheless, the time was a worrying one for Beauvoir.

What happened when Sartre did return to Paris signaled the most serious threat to the couple's complex relationship that had yet arisen. Sartre talked incessantly of Dolorès; he stressed his absolute compatibility with her. He announced his intention to spend a quarter of every year with her. Beauvoir, who was not afraid of Sartre's simple absence in itself (after all, they had led semidetached lives for much of their association), was, at first, uneasy, then alarmed. She worried that Sartre and Dolorès might share a depth of personal harmony that overrode her fifteen-year "understanding" with him. She forced a showdown by asking, directly, "'Frankly, who means the most to you [Dolorès] or me?'" Sartre's answer was equivocal: "'[Dolorès] means an enormous amount to me, but I am with you.'" Beauvoir said that his answer took her breath away. He clarified himself later, reminding her that they "had always taken actions to be more truthful than words, and that is why, instead of launching into a long explanation, he had invoked the evidence of a simple fact." Beauvoir said simply, "I believed him."[24] Beauvoir's statement of faith in Sartre looks doubtful and forced, but it was clearly the best assurance of his commitment she could get at the time. Some of the reasons for Sartre's uneasiness, given his immense unknown debt to Beauvoir, are obvious, but it is also true that neither he nor Beauvoir – despite Bost, despite Dolorès – wanted to let the other go. Beauvoir and Sartre's connection held together, if more tenuously than previously. For all that, Sartre was still wild to return to America and to his new lover, and he badgered all his contacts in Paris, trying to arrange for his return, which was finally set for December.

Before Sartre travelled back to the States, his and Beauvoir's status underwent a magnitude of change that both outdid their headiest dreams and fixed them permanently as public figures. The fact that what Beauvoir called the "Existentialist offensive" should be launched with such extraordinary success just as the pair's private relationship was under considerable strain has a certain irony.[25] The pair was touted as one of the exemplary couples of the century just when there was a good deal of doubt as to whether they would remain a partnership at all.

The shape of their lives to come was foreshadowed in a farcical incident in 1945. After de Gaulle had appointed Malraux as Minister of Culture,

and Raymond Aron as his Undersecretary of State, there was an attempt
by the couple's eccentric, homosexual friend, Marc Zuorro, to secure the
légion d'honneur for Sartre with the intention of embarrassing him. Sartre's
dislike for official honours was well known (he would refuse the Nobel
Prize in October 1964). Zuorro went to Sartre's mother and convinced her
to sign the papers Sartre should have endorsed to indicate his acceptance
of the award. "She thought," explained Beauvoir, "it was all very fine."
"The poor woman," said Sartre, "knew nothing about it at all. Her father
had had the *légion d'honneur*, her husband had the *légion d'honneur* . . . ."
Sartre said he "howled" when appraised of his imminent receipt of the
award. He had to appeal to Aron, who took his refusal amiss, to get the
proceedings cancelled. There was bad feeling all around. Although Sartre
did accept a prize associated with the Resistance (though his stumbling
comments – "I had it – Though God knows that my resistance . . . I was a
resister, and I knew resisters, but I never suffered much for it . . . I looked
upon myself not as personally worthy of this distinction, but worthy in
so far as other writers might, like me, have had the award . . . I stood
for a kind of French intellectual resistance" – indicate that he regarded it
more as an historical gesture than a personal honor), his general disdain
for such awards signaled his dislike for the sort of hierarchies implied by
them and for the intellectual mortality implied by their air of finality. As
Sartre explained in 1974 to Beauvoir, such awards also failed to meet his
own high opinion of his work:

> I can't see who has the right to give Kant or Descartes or Goethe a prize
> which means *Now you belong in a classification. We have turned literature into
> a graduated reality and in that literature you occupy such and such a rank.* I reject
> the possibility of doing that, and therefore I reject all honors.

Sartre had, as has been shown, other reasons for uneasiness with the
acceptance of honors for his work. And his explanation of his refusals are
particularly Sartrean in cast: half-charming and half-egomaniacal. Clearly,
whatever his position with regard to glittering prizes, Sartre wanted badly
to rank himself with the greatest of Western thinkers. In a statement rather
typical in its balance of truth and self-aggrandizement, he noted, finally,
to Beauvoir, "My deep reality is above honors."[26]

If the man who said he formed his desire for intellectual immortality
as a child in his grandfather's study was not about to be lured by the
awards bestowed by his all-too-mortal contemporaries who wished to
limit his reputation within their own era, still less did he welcome what
Beauvoir called "the inane glory" conferred on him (and less directly, on
her) by the existentialist offensive of 1945. Suddenly, the couple's public
literary productivity made them appear ubiquitous and prolific beyond

even their accelerating standards. In September alone Gallimard published *The Age of Reason*, *The Reprieve*, and Beauvoir's highly topical Resistance novel, *The Blood of Others*. The first issue of *Les Temps modernes*, with Sartre's impassioned editorial manifesto (dedicated to Dolorès), making the case for the need for *la littérature engagée*, appeared on October 15.[27] Sartre's famous lecture, "Existentialism is a humanism" was delivered on October 24 at the Club Maintenant. The room was so crowded that people fainted and Sartre could scarcely make himself heard. He wanted mass success and he got more than he bargained for. The couple were, said Beauvoir, "astonished by the furore" they caused. They were pursued by the press; photographed incessantly; they became the focus of gossip of all kinds. *Life* magazine did a spread on existentialism. As Beauvoir noted, "Sartre was hurled brutally into the arena of celebrity, and my name was associated with his."[28] Whatever their private difficulties, publicly they were a single entity. Sartre was both "a celebrity and a scandal." He refused to change his eccentric, donnish habits. He did not care how he dressed. He attempted to continue his life of living in hotels and working in cafés, and kept stubbornly to these routines long after they became impossible hindrances to him. The press was fascinated and appalled. Sartre's habits were presented as romantic and inexplicable. His and Beauvoir's relationship was another puzzle; they were not married, yet they were both too independent of each other to fit into recognizable patterns associated with "free love." Worst of all, said Beauvoir, the flood of media hype in the late 1940s violated Sartre's sense of himself as an aspiring immortal:

> He received worldwide and unexpected attention, but saw himself robbed of that of future generations. Eternity had collapsed . . . As a child, as an adolescent, Sartre's favourite fantasy had always been that of the *poète maudit* misunderstood by all during his lifetime and struck by fame's lightning only beyond the grave, or perhaps, so he can enjoy it a little, on his deathbed . . . His success had now overflowed all expectations, by accepting that loss, he nourished the secret hope that everything would be restored.[29]

It is with all this faintly ludicrous – but personally definitive – triumph, disappointment, paradox, and hope that Sartre embraced his era as fame took him by the throat. Beauvoir had always been attuned to her times. Her experience of the war had forced her to expand her attentiveness to the world beyond the self and out to the collectivity. Now Sartre did the same, not just in theory, but in earnest. With his typical energy, he not only embraced his role as a key voice for his time, but made the concepts of intellectual commitment and engagement peculiarly his own. As Germaine Brée says, the vocabulary Beauvoir and Sartre developed "of situation, commitment, choice, responsibility, guilt, and freedom" became the vocabulary within which "people did their soul-searching" from 1950 into the 1970s.[30] The impact the couple made on their world was immeasurable.

What astounded Beauvoir and Sartre was the amount of invective heaped on them along with the "inane glory." They were accused of immeasurable vice and of a degree of idiosyncrasy of behavior that even their eccentric arrangements did not begin to warrant. Sartre, in the end, seems to have accepted all this with some amusement. "'The advantage of our position,'" he remarked to Beauvoir, "'is that we can do whatever we want: it will never be worse than what they say we do.'"[31] Beauvoir's reaction to notoriety was both more circumspect and more profound. Always given to deception with people, and, as a woman, likely to receive less sympathy and understanding if the true complexities of her personal life ever became widely known, Beauvoir seems to have embarked on a systematic laying of false trails about herself, Sartre, and their friends which would continue to structure her biographical reportage of Sartre and autobiographical accounts of herself until her death.[32] By now she had served her literary apprenticeship and understood intimately the novelist's craft. She clearly decided, from the outset of the couple's fame, to mold her public commentary on their lives into classic narrative patterns that would make sense to their multitude of unknown followers, give some measure of the truth about their circumstance, and yet deflect attention from the more dangerous aspects of their lives. Part of this project included the glorification of Sartre, the diminishing of her own intellectual skill, and the suppression of information regarding her part in the development of their philosophy. If Beauvoir never told any story that was not part of the public record quite straight, she also left clues littered through her reports that indicated directions of meaning running counter to her overt statements. But more than anything, she promoted a myth of her and Sartre's association better and more intriguing than any less skilled publicist could have conceived. Beauvoir insisted repeatedly that she always meant to make her own life into art: in the end, she made it not only into art but into history. What Sartre thought of this project, or how much of it he came to believe, but with which he cooperated and which was probably a crucial factor in keeping the couple inexorably intertwined, not only during their lifetimes but beyond, is not known.

That some part of Beauvoir and Sartre's relationship, at least in Beauvoir's eyes, still consisted of the friendly competition of two star students is indicated by some of her statements about her reactions to Sartre's greater fame during the existentialist heyday. In the autumn of 1945, in the midst of first being thrust into the limelight, she considered their relative positions and declared to herself that she was not envious: "my first book had only come out two years before, it was not yet time to start adding up scores." That the "scores" would someday be added up, she had no doubt. A part of Beauvoir craved the kind of contemporary attention Sartre mistrusted, and she wanted it in her own right and not as part of a couple. "I wanted," she recalled after she had achieved her aim, "to be widely read in my lifetime, to be esteemed, to be loved. Posterity I didn't give a damn for. Or I almost didn't." With the full scope of her achievement hidden, with her own connivance, and without Sartre's mythomania about great writers living exclusively in an Eternal Sphere, Beauvoir took the existentialist triumph both more and less seriously than her companion. Cynically, she saw their fame as part of a French cultural offensive. She noted that, after the war, "now a second-class power, France was exalting her most characteristic national products on the export market: *haute couture* and literature." In this light, it is easily understood both how amused and offended Sartre and Beauvoir were to be lumped in with "existentialist" fashions for turtleneck jumpers and dive bars. More hopefully, Beauvoir felt that her current fame, though slender in comparison to Sartre's, opened up possibilities for work and success for her that were still to come. She was, she recalled, "satisfied but not satiated."[33]

Some of the unalloyed satisfactions of Sartre and Beauvoir's public position came through new personal contacts. While the family would never be abandoned by the couple, their circles of acquaintance widened even more in the period following the war. Most of their previous associations had depended either on their privileged education – which meant they were known, if only casually, to the nucleus of the French intellectual elite of their generation since their student days – or on their theatrical contacts, which derived from Sartre's first lover, Simone Jollivet, and her later partner, Charles Dullin, who had been friendly since the 1930s. The family, too, brought new contacts, though these tended to consist of Sartre and Beauvoir's former students. Now these networks remained in place while others were added. From 1945 onwards, the couple could meet whomever they wanted. One of their first post-war celebrity encounters was arranged by Nathalie, who had struck up a friendship with an American soldier who turned out to be Hemingway's brother. When the American novelist turned up at the Ritz as a war correspondent at the time of the Liberation, Sartre and Beauvoir joined him in his room for a night of enthusiastic drinking. Sartre staggered home at about 3 a.m.: Beauvoir stayed till dawn.[34]

In this period, too, Sartre renewed his contact with Nathalie Sarraute, who had sent him a copy of *Tropisms* before the war and sought out his acquaintance. Beauvoir met her in 1945 and spent a good deal of time with her over the winter. Violette Leduc, who developed a somewhat annoying passion for Beauvoir, was introduced to her in a cinema queue. Beauvoir, who would write the introduction to Leduc's *La Bâtarde* twenty years later, took the disturbed and disturbing lesbian writer into her life. They had dinner together twice a month for years. Leduc's physical, emotional, and professional claims on Beauvoir were so urgently pressing that Beauvoir had difficulty keeping her at a tolerable distance. Her friends often advised her to break off the connection, but Beauvoir, with her own intense standards of loyalty, refused, and stood by Leduc through all her long years of obscurity.[35]

In 1945 Beauvoir could not be certain that, despite her current fame, she would not sink back into obscurity herself. Her play, *The Useless Mouths*, opened on October 29 at the Théâtre des Carrefours, with Olga in a starring role. Beauvoir, unfortunately, sat next to Genet during the first performance, who kept up a running commentary on the work's inadequacies.[36] The play was not a great success, but it did run for fifty performances before it was taken off, and Sartre highlighted it as a significant example of the new French theater in a lecture in New York in 1946.[37] Beauvoir dedicated her only play to her mother, who was pleased by the gesture, and also designated one performance as a benefit for orphans of war deportees. Unhappy that theatrical success had evaded her, yet pleased enough with the modest gains it brought, Beauvoir's theatrical venture was more significant than she thought at the time. Its themes of responsibility to the marginalized members of the collectivity would lead directly into her work both on *The Ethics of Ambiguity* and on *The Second Sex*. The material effects of moral choice was one of Beauvoir's central interests; the play pointed the directions she was to follow in her analysis of this connection.

On December 11, 1945, Beauvoir gave her own lecture to the Club Maintenant. On December 12, Sartre boarded ship for New York: he would be in America until April, lecturing in both the United States and Canada. His attachment to Dolorès was now so strong that he contemplated taking a post at Columbia University and marrying her.[38] Again, his passion for Dolorès led to a change in his travel plans. He had intended to return to Paris at the end of February: in the end, he did not come back until two months later. During this time he seems to have written little: he was engrossed by his love affair, by his American lionization, and by New York.

Beauvoir was left in Paris to carry on with *Les Temps modernes*, which, though invariably spoken of as Sartre's journal, was for long periods of time more her project than his. By her own account, this time was a misery for her. Often in tears with her anxiety about her relationship with

Sartre, surrounded by people who made incessant professional demands on her in connection with the journal, but lacking the intimate contact with the family she needed (for part of this period Olga, who became worryingly ill after the failure of *The Useless Mouths*, Bost, and Camus, as well as Sartre, were all simultaneously out of town, while Nathalie had left permanently for the States with her new American husband, Ivan Moffat), Beauvoir simply got on as best she could during a period of depression. Physical conditions were still not good. Food remained short, she never had enough to eat. Her room was too cold to work in, yet the Flore, which provided such a refuge during the war years, had become impossible as a working base because of her fame. Beauvoir was reduced to writing in the basement bar of the Pont-Royal, which had casks for tables, which made work difficult. Beauvoir (like Sartre) was drinking heavily, and, given these circumstances, it is scarcely surprising that she literally jumped at the chance to go on a lecture tour of Tunis and Algeria for the Alliance Française at the end of January 1946.

Beauvoir had hoped to make her first plane flight with Sartre: it was one of their long-term joint fantasies. As she flew, for the first time, alone, to Tunis, she determined to infuse the journey with a compensatory glamor of adventure. When no one met her at the airport, Beauvoir, penniless, decided all the more to read her situation in terms of "unexpected freedom."[39] As she had during her years in Marseilles, Beauvoir took to solitary walking expeditions with a show of bravado and fecklessness. (During one lonely walk in the dunes, she awoke from a nap to find "a very dirty old Arab" sitting on her stomach. She escaped his half-hearted attack after emptying her purse into his hands.) The trip was a welcome interlude for Beauvoir, but Sartre – whom she expected to return – still was not back in Paris at the end of her African lecture tour. Beauvoir had finished her third novel, *All Men are Mortal*, at the beginning of 1946. This phantasmagoric allegory, which owes something to Virginia Woolf's *Orlando*, something to surrealism, and more to Beauvoir's historical research in the Bibliothèque Nationale, played out her meditations on death and immortality on an epic scale. The novel, published in November of 1946 – like its related production, *The Useless Mouths* – failed to attain the popularity Beauvoir thought it merited. The work she took up in 1946, *The Ethics of Ambiguity*, Beauvoir herself consistently underrated. Her project was to construct an existentialist ethics. She describes the genesis of the book as a response to the challenge to "base a morality on *Being and Nothingness*", a task she was eminently suited to perform.[40] One of Sartre's former students and Camus, she said, were instrumental in encouraging her to undertake the project. In her own retrospective critique, she argues that the work is undermined by its abstraction, by its Hegelian categories, and by its Kantian absolutes.

Her error, she felt, was to pay insufficient attention to the historical placement of ethical choices. And while her hope had been to rescue existentialism from charges of nihilism and pessimism, her defense, she thought, had been inadequate.

Beauvoir, however, misrepresented her work when she castigated *The Ethics of Ambiguity* for its historical blindness, as she did her contemporaneous moral essays for *Les Temps modernes*, which were collected and published in 1948 as *L'Existentialisme et la sagesse des nations*, for excessive philosophical idealism. As usual, she draws attention away from her philosophical achievement in what were, in fact, crucial works in the second phase of development of her concept of the Other. Now that Beauvoir and Sartre's primary interest had shifted from the individual to the social, they urgently needed to expand their philosophical system in order to discuss their new concerns with the same profundity as they had their old. Once again, the task of initial deep thinking fell primarily to Beauvoir, and, this time, so did the task of initial articulation of her philosophical ideas in essay form, a skill for which she lacked Sartre's easy facility. Her theory of the Social Other would not emerge fully until the appearance of *The Second Sex* in 1949, but in her *Ethics* she made considerable headway in its development. In addition to their need to invent an existentialist ethics, Beauvoir and Sartre wanted to devise a system of analysis of social oppression that was neither Marxian nor Freudian, and one that was founded on individual existence. Beauvoir's response to this need was to prove one of the decisive turns in twentieth-century social philosophy. Sartre, unsurprisingly, adopted her theory of the Social Other as his own. Frantz Fanon, the anti-colonialist philosopher and social commentator, was, in turn, to adopt it from Sartre in ways that had striking effects during the period of decolonization. Most importantly, her new theory was to provide Beauvoir with a mechanism through which she could account for the oppression of women, just as Darwin's theory of natural selection had provided the causal mechanism of his theory of evolution. The revolution in possibilities for analysis are directly comparable: Beauvoir must be judged one of the pivotal thinkers of her century.[41]

The Ethics of Ambiguity, while hastily constructed, and despite Beauvoir's criticisms, is thus notable for its intense interest in the historical and the social. Although, as Beauvoir says, in it the individual is still the centre of her attention, that individual is located within social roles not of their making. With striking originality, and in contradistinction to Sartre, but very much in keeping with her own experience (and with the thrust of her fiction), Beauvoir identifies adolescence as the crucial period for the formation of the moral self. It is at this juncture between childhood and maturity that the individual constructs their idea of self and other, and their relationship to the roles intended for them in their social context. Beauvoir, in anticipation of her future work, particularly stresses the situations of the colonized, of women, and of children as primary examples of those who are intended to be controlled and infantilized as Social Others in contrast to the roles of powerful adult male colonizers, who claim the right to make laws and determine truth. Beauvoir underlines the need for the oppressed themselves to seize their chances for liberation. The oppressed, she writes, must assume their moral status, must assume their subjectivity: "once there appears a possibility of liberation, it is resignation of freedom not to exploit the possibility, a resignation which implies dishonesty and which is a positive fault."[42] This was a position Beauvoir was never to abandon, however much she might modify her accounts of the difficulties for the oppressed of assuming responsibility for their own liberation. Beauvoir had, indeed, absorbed intellectually the historical lessons of the war. Her new ethics reached out to the social world – and with an unshakeable if stern optimism about the self-determining powers open to the oppressed for which some of her critics have still not forgiven her. And if she later felt that her materialism was not far enough developed for her to articulate it properly in her writing at the time (contemporary quarrels with Aron and Maheu indicated, she said, her passionate belief in the utterly fundamental importance of the means of life over abstract principles), her work in 1946 nevertheless laid the foundations for the intellectual and political paths she was to follow.[43]

While Beauvoir was wrestling with the construction of a new ethics which would point the direction for both her and Sartre's subsequent work, Sartre completed his second American interlude. He was back in Paris in April and seems to have begun, almost immediately, a new contingent affair with Michelle Vian, the wife of jazz trumpeter and writer Boris Vian. Beauvoir had become friendly with the couple in his absence, and Sartre's annexation of another sexual partner signaled a diminishing of his obsession with Dolorès.[44] Michelle, like Olga and Wanda – other women that Beauvoir, in various ways, "shared" with Sartre – was to become a fixture in the couple's daily lives, and was one of the many dependants Sartre would support financially for years. The provenance of Beauvoir and Sartre's relationship was still in jeopardy, despite Beauvoir's rather enthusiastic acceptance of Michelle. The "primary" couple had not slept together for some years, but their intellectual partnership was as crucial for each other as ever. In 1946 Beauvoir was seeing Maheu and Bost; Sartre, recovering from a bout of mumps – nursed by Beauvoir – was seeing Michelle and Wanda. Things seemed to be returning, if not to normal, then to something that resembled the couple's established patterns of complex intimate behavior.

However, 1946 brought significant changes as well. For one thing, Sartre was now rich. He treated his wealth both with characteristic generosity – he continually assumed responsibility for the financial support of his friends and associates – and with a certain ironic, clownish swagger. Sartre took to carrying rolls of notes in his pockets ("'It makes me look like a big shot to pull out a bundle,'" he commented in an interview), insisted on paying for everyone in restaurants, and tipped waiters with mad extravagance.[45] Sartre hired his first secretary, the young Jean Cau – who would stay in his post for the next eleven years – in June. In October, Sartre's living situation changed dramatically when he moved into his mother's flat at 42 rue Bonaparte, his address for the next sixteen years. Given the best room as a study-bedroom, Sartre, for the first time, built up a working library.[46] His mother, who treated Sartre with the respect she had reserved for Mancy, referred to their cohabitation as her "third marriage."[47] The piano in the flat provided an outlet for Sartre's musical abilities, and he played, on most days, sometimes with his mother, and sometimes with Beauvoir, for two hours in the afternoon.[48] In his 41st year Sartre returned to live with his childhood room-mate. His mother's treachery in marrying Mancy had, it seems, been forgiven.

Whatever the rather dubious psychodynamics of Sartre's return to his mother's home, his new residence provided him with a place to work now that the cafés had become intolerable for the chief existentialist celebrity. If the initial stages of Sartre's liaison with Dolorès seems to have impaired his ability to work, he now kicked into high gear. The physical regime Sartre adopted to support his prodigious output was punishing. "Coffee to stay awake, orthedrine pills to keep the speed up, whisky to relax" – Sartre kept himself writing by whatever means gave results.[49] In their youth, he and Nizan had decided they were "supermen": for the next years Sartre undertook to prove that myth with regard to himself.[50] The remainder of 1946 saw Sartre on lecture tours in Switzerland and Italy. His theatrical campaign continued with the premieres of *Men without Shadows* and *The Respectable Prostitute* in November. He published "Materialism and revolution" in *Les Temps modernes* in June and July, and worked on a screenplay for *No Exit* during a holiday in Rome (which was to become a welcome summer habit for Sartre and Beauvoir). He wrote articles on America, literature, Descartes, art, and politics. Indeed, his political interests were again deepening. *Existentialism is a Humanism* had been published in January; *Anti-Semite and Jew* appeared in November; *Baudelaire* would come out early in 1947. Sartre began work on *What is Literature?* after a trip to Holland with Beauvoir in the autumn.[51] Sartre's energy and productivity during this time were amazing, the provocativeness and excitement of his writing undeniable. Sartre had indeed thrown himself into his era and there were few aspects of his time on which he felt he could not speak. With Beauvoir's help he made himself the public voice of independent, left-wing, post-war France.

If Sartre had made himself into the preeminent spokesman for his time and place, 1947 was the year in which Beauvoir would begin her work on women that would reach out definitively beyond its historical boundaries.

The end of 1946 had been discouraging for her. In late summer, before the trip to Rome with Sartre, she made what she said she knew in her heart was her last solo walking tour. Beauvoir enjoyed her three weeks in the Dolomites in a bittersweet mood.[52] She and Sartre met Arthur Koestler in October: Beauvoir slept with him once, then decided he was loathsome.[53] There were fallings out with Camus, who was irritated by the continuing existentialist craze. *All Men are Mortal*, on which Beauvoir pinned high hopes, came out in autumn and attracted hostile reviews. It failed to sell. Dolorès was coming to Paris early in 1947. Beauvoir badly needed to have her spirits lifted. She grabbed at her upcoming first trip to America, arranged as a lecture tour by Philippe Soupault for the Alliance Française, as a means to "feed" her happiness.[54] This transcontinental tour was to yield a rich load. It produced her study of America, *America Day by Day*, and a number of substantial journal articles. It gave her the passionate love affair she was to mythologize as thoroughly as she did her primary relationship with Sartre. It provided a lever to use against Dolorès. And it gave the final impetus for the composition of *The Second Sex*.

The first stages of her 1947 American tour, which lasted from the end of January until May 18, gave no indication of just how successful this excursion was to become. In fact, it looked as if the pattern of misfortunes was continuing. Beauvoir's flight from Paris was worryingly delayed.[55] When she arrived in New York the finances of her tour had not yet been settled and she had to rely on the good offices of her old fellow student (and Sartre's cousin) Claude Lévi-Strauss to sort out her difficulties. With her usual physical stamina, Beauvoir tramped the streets of New York with excitement, delight and an utter sense of freedom. She was fascinated by drugstores, and took to American food and alcohol with pleasure. However, at first she had difficulty communicating: though her English was excellent, her accent was heavy, and she made slow headway in making herself understood. In her official capacity as existentialist ambassadress, she found herself annoyingly caught between American interviewers, who tended to neutralize her as an elegant Parisienne, and the conservative French colony, which was composed largely of the sorts of people for whom she (and Sartre) felt complete antipathy (which was, of course, returned by her hosts who found her something of a scandal at the same time as they lionized her).

To her shock, Beauvoir found herself reading Americans in the ways Sartre had suggested to her. In the mass, and on their home ground, Beauvoir thought Americans inert, dull and filled with a nationalistic chauvinism which, she said in a most damning comparison, would have been worthy of her father. She cites William H. Whyte's later *Organization Man* as the appropriate cognate American source for her impressions of America and its inhabitants. She was particularly disappointed by American women. They were not what she had expected and they did not evince the freedom of action and mind she wanted to find. They were still, it seemed to Beauvoir, "dependent and relative" beings.[56] Along with this unexpected sexism, she was appalled by American racism (though she liked the American South in spite of it, particularly New Orleans with its predominantly French influences), and she made racism one of the focal themes of *America Day by Day*, in defiance of members of the American French community, who made a point of warning her off the subject.[57]

Aside from her public duties – lectures which Beauvoir soon found ridiculous, given her usual audiences of docile, conformist college girls, superficial culture addicts, and scandalized academic French expatriates – her personal contacts were also, at times, disappointing. Although she was clearly interested in Lévi-Strauss's research, and used some elements of his structuralist anthropology in fascinating ways in *The Second Sex*, she found him and his wife tiresome. Similarly, the *Partisan Review* crowd, which courted her and which was turning to the right under the pressure of the growing Cold War, filled her with disgust. Little of Beauvoir's writing had yet been translated into English, so she tended to be judged solely as Sartre's disciple and companion. She found her prospective American editors demanding and unreasonable. Existentialism, indeed, French intellectual life in general, seemed to her to be treated in commercial terms alone. "They talk about existentialism," she reported to Sartre, "as they would about a worm powder, and they award marks; Camus so many, Sartre so many – my God! by the end I was quite scared."[58] She was increasingly put out when Bost did not write.

The annoyances of this trip seemed unending. One of Beauvoir's first "social" duties when she arrived in New York was to meet Dolorès, who was about to embark for France. Beauvoir was careful to write to Sartre about how charming she found his choice of partner, but also took the opportunity to describe Dolorès's heavy drinking, which resulted, she said, "in a certain nervousness, a certain volubility, and some classic crazy behaviour."[59] Stépha Gerassi, who was present at the rivals' meeting, found the occasion puzzling. Beauvoir did not react to the meeting as her friend had expected. " 'It was as if this woman was her obsession, but once she saw her, well, that's that and it was all over.' "[60] Beauvoir sized up her rival correctly (and, after all, she had lots of experience in this sort of thing). She seems to have stopped worrying about Dolorès and set about restoring the balance in her and Sartre's sexual ledgers. In this she was to succeed beautifully.

There were people whom Beauvoir enjoyed seeing in America. The Gerassis, in New York, were as sympathetic as before, and she particularly liked their son, John, who would become one of Sartre's biographers. In Los Angeles she took great pleasure in staying with Nathalie Sorokine, who travelled with her through the American West, and whose husband, Ivan Moffat, a film producer, caused her much excitement by proposing a film of *All Men are Mortal*. Beauvoir also liked the black novelist Richard Wright and his wife, whom she had met in Paris.

But Beauvoir's most significant connection in America was with Wright's fellow Chicago School novelist, Nelson Algren, whom Beauvoir ran to ground on the strength of an introduction from one of Algren's former lovers. Algren, who was to win the American National Book Award with the novel he was then completing, *The Man with the Golden Arm*, was a godsend for Beauvoir.[61] She annexed him at precisely the right time, when she needed a new and serious love affair to counter Sartre's with Dolorès, and when both Sartre and Bost were failing her in significant ways. As much as Dolorès had given Sartre America, so Algren gave the country to Beauvoir. He intrigued her: he was both completely comprehensible and a total mystery. He, like her, was left-wing, and congenitally anti-bourgeois in his opinions and in his life. He lived in a way Beauvoir recognized, in "his little, indigent intellectual's room" in a manner she understood.[62] She shared Algren's taste for alcohol and the seedier sides of cities. The physical passion of their affair was welcome (and portrayed convincingly in her famous fictionalized account of their relationship in *The Mandarins*), and Beauvoir tended to look at it as her last renewal of bodily youth before the acceptance of the ageing she dreaded. The affair was to end only in 1950 after hundreds of letters and several visits between the two lovers. Algren was loath to believe that Beauvoir would not leave her life in Paris, and sacrifice her connection with Sartre for him. He remained bitter about what he saw as her betrayal of him and of the confidentiality of their lovemaking throughout his life, and was still speaking angrily about her a few days before his death in 1980. For her part, Beauvoir always wore, and was buried wearing, the silver ring Algren gave her in 1947.[63]

Beauvoir portrayed her love for Algren as the fruit of irresistible mutual passion, but this seems to have been only part of the story. Much has been made in recent years of Sartre's sexual rapaciousness, calculation, and double-dealing. In this, as in so many other aspects of their union, Beauvoir was more his match than is usually suggested. Algren was undeniably useful to Beauvoir in a number of ways. She had been angered when Sartre asked her to delay her departure from America for ten days at the beginning of May so that Dolorès could extend her time in France. Beauvoir immediately responded by heading straight back to Algren in Chicago, and let Sartre know it. When she returned, later in the month, to Paris, she and Sartre were soon hiding from Dolorès and her serious demands on Sartre, in a cottage on the edge of Paris, with members of the family, who saw their own privileged positions undermined by the interloper, cordoning off the couple who underwrote their lives. Dolorès went back to America somewhat baffled: eventually she gave up on Sartre. For his part, Sartre clearly took fright at the seriousness of the counterthreat Algren posed. Beauvoir was not only his companion, she was his reader, his mainstay at *Les Temps modernes*, the hidden source of some of his most important ideas, and, finally, at least his equal in the areas of life that counted most for him. He could not afford to lose her; Beauvoir did not wish to be lost. One almost, in retrospect, feels pity for Nelson Algren and Dolorès Vanetti. Once again, Sartre and Beauvoir accommodated themselves to each other: their partnership had held together. They settled down to enacting their public role as Sartre-and-Beauvoir, a role that would last until the end of their lives.

EPILOGUE: SITUATIONS IN MATURITY

Whatever the tensions and stresses in their lifelong relationship, what is utterly clear about Beauvoir and Sartre is that they chose each other, repeatedly, as beloved friends, companions, collaborators, and *alter egos* over all other possibilities (and there were many) at every juncture of their careers. Tracing the details of those careers from the late 1940s to the end of their lives will be the focus of the projected second volume of this biography. What can be done here is to indicate the impact on Sartre and Beauvoir of some of their major commitments and projects in their highly productive years of maturity.

Rassemblement Démocratique Révolutionnaire

In his interviews with Beauvoir in 1974, Sartre noted the high place politics held in his anticipated construction of himself as a "great man." He recalled his youthful admiration for Victor Hugo, Zola, and Chateaubriand, and explained his political ambitions in terms of their histories:

> Those lives formed a synthesis, producing a single life that was to be mine. I really behaved according to these patterns, and I thought I should dabble in politics when I was fifty . . . I didn't think politics was life, but in my future there had to be a political period.[1]

Sartre's partly self-mocking, partly grandiose description of his political ambitions wryly (and somewhat bitterly) undercuts the passion and sincerity that he brought to the political roles he did, in fact, play

168

during his life. His excursion into an attempt to form a national political movement – that both was and was not a political party – which was intended to cut through France's increasingly polarized divisions into right and left in the years 1948 and 1949, demonstrates both the nature of his political commitment and the disappointment to which that commitment was subject.

The events of the late 1940s brought widespread dismay to many former members of the Resistance who hoped to shape a new, idealistic political order in the aftermath of victory. The widening political fissures in France were confirmed when de Gaulle left his post as president of the provisional government in January 1946. The post-war consensus, which had held together in the immediate, comradely euphoria of the Liberation, was breaking down quickly. In October 1947 de Gaulle, now firmly associated with the right, returned as a political force after the victory of his newly formed Rassemblement du Peuple Français in the municipal elections, which were held against a background of political instability fueled by waves of strikes by the press and by the transport workers, and by the proposals for the Marshall Plan, made in June 1947, which accelerated the division of Europe into pro-Soviet and pro-American blocs.[2] Disillusioned by the return to (and intensification of) pre-war political configurations, and appalled by the terror instilled into the population at large by the new Cold War, Sartre used his editorial control at *Les Temps modernes* to argue for a non-communist and democratic socialist position, grounded on Merleau-Ponty's ideas of a humanist Marxism. On the day after the first of the Gaullist election victories in October 1947, a radio series, *La Tribune des Temps modernes*, featuring Beauvoir, Sartre, Merleau-Ponty, and other members of the journal's staff, made its initial broadcast. The series, which attacked Gaullism and the emergent Cold War, attracted widespread interest and angry criticism. Significantly (and painfully), it prompted a quarrel between Sartre and his old friend, Raymond Aron, which was to rupture their association for years. The *Tribune* series was suppressed after only six programs when the socialist Prime Minister, Paul Ramadier, was replaced by the leader of the Christian Socialists as the head of a coalition government which clamped down on free criticism of its policies.

Among the confusing and dispiriting realignment of post-war national and international political interests, members of the non-communist left in France felt their influence waning. Sartre understood the urgency of working to strengthen the middle ground in an era in which the fear of nuclear war was becoming ever more pervasive, and of promoting what he later defined as "the well-known third force that the socialists would have liked to be",[3] before the middle ground disappeared as an option altogether. In November 1947, he signed two socialist appeals for world peace. In February of 1948 he agreed to join the newly formed Rassemblement Démocratique Révolutionnaire (RDR), which tried to secure the possibility of that European third force to which Sartre was committed. The RDR opposed both capitalism and Stalinism. It devoted itself to building precisely the kind of non-communist socialism in Europe that had been the dream of many of the participants in the French Resistance. Sartre was a signatory of the RDR's manifesto which appeared in February 1948; and he, David Rousset (an ex-Trotskyite who had survived imprisonment in Buchenwald), and Georges Altman (the editor of the popular newspaper, *France-tireur*) became the most important figures associated with the movement. For over a year Sartre not only gave enormous amounts of time and energy to the RDR, he more or less bankrolled the experiment, donating 300,000 francs for its operations.[4]

Sartre's hopes for the RDR grew directly from his experience of the Resistance, which he tried to draw upon in the new political circumstances. The clarity of values which had underpinned Socialisme et Liberté was impossible to achieve in the scramble for post-war political ascendancy, but Sartre's vision was of a utopian order nonetheless. His contribution to the RDR was infused with a commitment to grass-roots control and organization, which foreshadowed similar structural reforms by the left in the late 1960s. In his notes for his plans for the RDR, Sartre's desire to find a complex synthesis for opposed political forces is strikingly evident. In *Force of Circumstance*, Beauvoir reproduces Sartre's working notes on his hopes for the RDR. He wants, he writes, the movement to unite the "middle classes and the proletariat," to create a third force – "Europe. Not America, not the U.S.S.R., but the intermediary between them." He wishes to secure both "democratic and material liberties."[5] In all this, writers and intellectuals were to play key roles. This is a new version of the old dream of a writer's utopia and, indeed, Sartre's own contemporary work on the concept of engaged literature, which was to be one of his most important popular conceptualizations of the post-war period, is crucial here. Again, as his working notes in the late 1940s reveal, the connection between literature and politics was strong in his own mind at the time: "*What is Literature?*," he writes, "led me into the R.D.R."[6]

The Rassemblement Démocratique Révolutionnaire held together only briefly, though its short life was indeed one of hope for the recurring dream of uniting the non-communist left as the holder of the central impetus in European politics. Sartre left the RDR in October 1949, after months of disillusionment with the power-struggles within the organization, and after quarrels about the developing anti-communist policies that seemed to him to be dragging the movement to the right. For Sartre, his withdrawal was a grave matter: "'Splitting up of the R.D.R.',," he wrote in his contemporary notes, "'Hard blow. Fresh and definitive apprenticeship to realism. One cannot create a movement.'" Beauvoir, who had watched the rise of the RDR with interest and supported its development, also felt defeated.[7] In 1974, Sartre characterized his political life from 1949 until 1968 bleakly as one of "total solitude."[8] This was not quite true, but until Sartre thought that he rediscovered in Maoism in the late 1960s some of the principles that had motivated his support for the RDR in the late 1940s, both he and Beauvoir were finished with political parties and general political movements. From this time on they gave their support to issues rather than to organizations. They accomplished a good deal, but Sartre seemed to have judged his post-war political experience, indeed his political effectiveness in general, as an exercise in abjection.

The Second Sex

Sartre's failure to achieve his stated political ends in the RDR did not mean, however, any diminution in his post-war opinion of the importance of politics itself. In this, he felt himself to be utterly unlike Beauvoir, who, he believed, was devoid of political aspirations. In an interview on Beauvoir in 1965, he revealed his lack of comprehension of Beauvoir's political strategies:

> It is only on one subject that she leaves me flat and that is politics. She doesn't give a damn about it. It's not that she actually doesn't give a damn about it, but she doesn't want to get involved in the political rat race.[9]

In some senses, Sartre's somewhat bemused comments are apt. Beauvoir was not particularly "political" in the traditional manner, in the way that Sartre, with his patterns of engaged male writers from the past as his templates, would have liked to be. But Beauvoir was a profoundly political animal for all that, although her politics were of a different order than the "rat race" variety Sartre mentions. Beauvoir's achievement in this area was to become one of the most important theorists of the politics of the Other – of the marginalized, the oppressed, the disenfranchised. While Sartre was busy with the manifestos, the in-fighting, and the financing of the RDR in the late 1940s, Beauvoir completed and published one of the landmark political books of the century, her monumental and multifaceted study of the condition of women, *The Second Sex*.

As befits a book of such significance, there is an extensive mythology surrounding the composition and the reception of *The Second Sex*, much of it shaped and presented with characteristic narrative flair by Beauvoir herself. In many ways, Beauvoir's study of women represents the direct and linear extension of her concept of the Social Other which she had been developing in her fiction, philosophy, and journalism since the composition of *She Came to Stay* in the late 1930s. The Beauvoirian Other provided the missing mechanism needed for an intellectually convincing analysis of social oppression whose origins were not primarily, or, at any rate, not solely, economic. In her analysis of the ideological roots of women's subjection, Beauvoir reaches far in advance of her historical position. In other ways, understandably, *The Second Sex* belongs deeply to its time and place: it is a universal study of women that is nevertheless focused on the experience of (primarily) bourgeois French women of the nineteenth and twentieth centuries. From yet another point of view, the book belongs inextricably to Beauvoir's own life, and when she came to write her memoirs, beginning in the 1950s, much of their structure and content was constructed with an exacting eye to the concepts that arrange women's lives as Social Others which Beauvoir understood profoundly after her completion of *The Second Sex*.[10]

Beauvoir said that she began work on *The Second Sex* in October 1946, immediately after completing *The Ethics of Ambiguity*. She finished the second volume of her study of women in June 1949, taking time out during its composition to write the lengthy, if journalistic, *America Day by Day*. Extracts from *The Second Sex* first appeared in *Les Temps modernes* in 1948 and 1949, pushing up sales of the journal and building a potential market for the book. Gallimard brought out the completed first volume in June 1948: it sold 22,000 copies in its first week.[11] The second volume appeared in November. Dealing with contemporary issues, in contradistinction to the first volume, which laid out Beauvoir's theoretical principles in the light of women's history, the second volume caused even more of a stir than the first. Beauvoir's study was singular for its time. It appeared in a period without a visible contemporary feminist movement. As a seemingly lone voice, Beauvoir found herself the subject of incredulous curiosity and of abuse both in the French press and in the streets of Paris. Even some of her friends, notably Mauriac and Camus, turned on her. The personal nature of the attacks appalled and outraged her. She was accused, she said, of being "Unsatisfied, frigid, priapic, nymphomaniac, lesbian, a hundred times aborted, I was everything, even an unmarried mother."[12] It must be recalled that this treatment was afforded to Beauvoir before her fame was decisively established, and before she had provided most of the documentary material which was to form the substance of the Sartre–Beauvoir legend. The impact of the reception of *The Second Sex* on Beauvoir's decisions about what and what not to include in her autobiographies and interviews cannot be calculated.

What can, however, be said, is just how important it was for Beauvoir to have the kind of public success that *The Second Sex* brought her at precisely the time she needed it. The book sold millions of copies around the world;

it was not only an instant classic, it was a best-seller. The royalties from this production released Beauvoir from financial dependency on Sartre, which had become slightly awkward after the readjustments in their relationship during the Vanetti/Algren affairs. It gave Beauvoir her own distinct public persona and freed her from the limiting public role of Sartre's companion and assistant. Most importantly, it reaffirmed her writerly confidence after the relative failure of her play, and after her misjudgment of the attractions of *All Men are Mortal*. Her minor, if significant, reputation as the high priestess of existentialism (which did not and could not take account of her primary role in the formulation of the movement's key philosophical ideas in the absence of information about her role in their construction), as adjunct to Sartre and his theories, now turned into a major reputation of her own. When Beauvoir finished *The Second Sex* she was ready to begin the novel which would win the Prix Goncourt, *The Mandarins*. In this production, too, the autobiographical impulse would be strong. Beauvoir was, indeed, putting her life to use, and, in so doing, her place in not only French but twentieth-century intellectual life was to be secured, if only partially understood.

Beauvoir had so internalized the material in *The Second Sex* that, by the time she wrote her book, she said that she felt more like the transcriber of its text than its originator. "'I said many things I deeply believed in,'" she said, "'but, in a certain sense, everything I wrote seemed natural to me, obvious: I was only the one who organized and wrote it, everybody already knew it.'"[13] The book loops back through Beauvoir's life in a number of ways Beauvoir was keen to highlight. Its origins are complex and include Colette Audry's stated wish to write such a study which she discussed with Beauvoir in their days as teachers in Rouen; Sartre's encouragement of Beauvoir to write about her life and about her experience as a woman; and Beauvoir's own sense of her perceived exceptionalism as a female intellectual when compared with the women she met in the early days of the existentialist offensive.[14] *The Second Sex* is complex, not only in its origins, but in the material upon which it draws. The argument takes note of a wide range of theoretical views (ranging from those of earlier feminist writers to Lévi-Strauss's anthropology and Lacan's new psychology), and is deeply inflected by Beauvoir's thoughts on equality during her time in America, which led her to consider the victimization attendant on race and sex as homologous.[15] But, overwhelmingly, the book is structured by the concept of the Beauvoirian Other, put to work with great success, at every level of the analysis of the history and place of women. Beauvoir was, unsurprisingly, pleased with the study and with its long-term effects. It solved a variety of problems for her. It was, she said in old age, "'possibly the book that has brought me the greatest satisfaction of all those I have written.'"[16]

In October 1948 Sartre's books had been put on the forbidden index by the Vatican (just as his work was later banned in the Soviet Union); *The Second Sex* and *The Mandarins* were similarly condemned by the Roman Catholic Church.[17] The ban helped sales enormously; it confirmed the couple's credibility as rebels even more. Beauvoir and Sartre had identified themselves, separately and undeniably, as voices of protest against a variety of established centres of power, as voices of rebellion with which the world had to reckon.

The Ambassadors

But as much as Beauvoir and Sartre were themselves rebels, and tended, in the main, to see themselves as such as long as they lived, their identities also included facets that were more amenable to cultural cooptation, which sometimes took forms that they themselves approved and which they promoted. In the wake of the existentialist offensive, the couple occupied the curious position of functioning both as darlings of the international left and bona fide French cultural icons. The contradictions involved in this dual role amused them, irritated them, and, at times, exasperated them, but it also brought opportunities from which they did not hesitate to benefit. While Beauvoir and Sartre remained unworldly and unconventional in terms of their shared disdain for the bourgeois obsession with accumulation of objects, property, and financial securities, they indulged their passion for travel liberally throughout the years of their full maturity. It was, along with dining out, their main extravagance. In addition to the summers spent in Italy (usually in Rome), which became an almost annual fixture in their calendars from the early 1950s on, Sartre and Beauvoir spent a great deal of time travelling in quasi-ambassadorial roles as representatives of France and as part of the international elite of the non-aligned left. Following their forays into the United States in the late 1940s, they made major trips to Africa (in 1950), to China (in 1955), to Cuba and Brazil (in 1960), to the Soviet Union (in 1962 and 1963), to Japan (in 1965 and 1966), and to the Middle East (in 1967), as well as making many shorter trips throughout Europe, which ranged from Greece to Iceland in the decades of their greatest fame. Both Beauvoir and Sartre moved with official ease back and forth across the Iron Curtain during the years in which it was most impenetrable. In addition to their joint travels, each made many trips on their own. The pair had always travelled light; now there were years in which they could truly be considered more as citizens of the world than citizens of Paris.

The photographs of many of their travels, and Beauvoir's accounts of their journeys in her diaries, make for odd reading. It is clear that, while the outlines of their global coverage are exciting and adventurous, much of the time the couple was locked into the box of official dinners and tours, conferences, and ceremonies – all the dreary paraphernalia of the ambassadorial role that is designed for the infliction of propaganda on visitors – that had so disgusted and bored Sartre on his first visit to the United States in 1945. Just as Sartre responded to his American boredom with the pursuit of a new lover, so his travels seemed to bring out his Don Juanism in the 1950s and 1960s. Like a cliché sailor, he wanted a different woman in every port of call, and Beauvoir occasionally had to help him out of self-inflicted difficulties, such as the one he encountered in South America in which he almost married the young Brazilian girl of the moment, who seems to have refused to sleep with him without being married.[18] Her relatives shared the same view and Sartre only escaped casually acquiring a Brazilian spouse with a good deal of effort. Beauvoir, on the contrary, had always been a passionate imbiber of facts and sights, and seems to have been the ideal recipient of whatever official views governments (especially of the left) wanted to promote. Beauvoir's book that was the result of the trips to China in 1955, *The Long March*, which does contain interesting moments, could, in the main, have been written by the Chinese propaganda ministry, and remains Beauvoir's most (and perhaps her only) embarrassing production. While Beauvoir seemed pleased with these years of hectic travel, and decorated her flat with souvenirs and mementos of her journeys, Sartre seems to have seen their value more abstractly from the first, and regarded each trip as a political act, and as an addition to the slate of activities that had composed his childhood dreams of greatness.[19]

Whatever their personal significance to Sartre and Beauvoir, their travelling was, indeed, notable for its political edge. They did what they could, by the very fact of their presence – a presence that was signaled to the world by the newspaper features and articles which followed their activities – to announce their allegiance to dialogue and diplomacy at a time when barriers to discussion between the communist and non-communist world grew ever higher, and the fear of nuclear holocaust ever greater. This politics of global presence was, without doubt, even more important than the humanitarian and libertarian meetings and conferences to which they often travelled to attend as speakers or delegates. Whatever Sartre's failure as a conventional politician, and whatever Beauvoir's stated general lack of political interests, the two understood the impact of the media on politics and used their fame to put it to work in the service of opening up the voids of silence and fear that were encouraged by both sides in the Cold War. In this endeavor, in which they used their fame both wisely and shrewdly, they stand forever pictured as a couple, serenely accompanying the leaders of both East and West, equally unafraid to criticize the West's barbarity in Vietnam or the East's intervention in Hungary and Czechoslovakia. As world citizens they functioned as they intended, not as icons of France, but as icons of intelligence, justice, and peace, working for dialogue in an historical period devoted to the politics of division and hysterical fear of the political other.

Algeria

The 1950s and 1960s were the great decades of colonial liberation, and the focal point for the process of decolonization in France was the bitter and vicious Algerian war of independence which fractured both countries during the protracted hostilities which lasted from 1954 until Algerian freedom in 1962. The war split the French political landscape into poisonous segments, carried de Gaulle back into power on quasi-dictatorial terms, and brought to the boil the racism and nationalistic chauvinism in France that the immediate aftermath of the Second World War had tended, somewhat, to obscure.

The Algerian war hit Beauvoir especially hard and affected both her writing and her overt political behavior in ways that were of crucial importance. *Les Temps modernes* came out openly on the side of the FLN (Front de Libération Nationale), the main organization of the Algerian insurrectionists, almost from the first. The editorial board was in constant receipt of documents detailing the routine French violation of human rights in Algeria, and particularly of the standard use by the French of terror, torture, and rape to subdue the Algerian population. Beauvoir reacted passionately to the situation. Her sleep was filled with nightmares throughout the years of the war, and she felt increasing disgust for her country and its people. In *Force of Circumstance* she says that she and Sartre began "by loathing a few men and a few factions; little by little we were made forcibly aware that all our fellow countrymen were accomplices in this crime and that we were exiles in our own country."[20] She felt that her "own situation with regard to my country, to the world, to myself, was completely altered" by the war and by what she knew about the French conduct in it and attitudes toward it. "I could," she said, "no longer bear my fellow citizens . . . I felt as dispossessed as I had when the Occupation began."[21] In response to the war Beauvoir did whatever she could: joined demonstrations, helped publish accounts of French atrocities, wrote articles condemning French action and in defense of political prisoners (an article she wrote for *Le Monde* on behalf of Djamila Boupacha, a young Algerian woman who had been raped and tortured by the French, led to the confiscation of that issue of the paper in Algiers), and acted as copresident for the anti-Gaullist committee for the 14th *arrondissement*. Her efforts seemed to lead nowhere. She and Sartre, who was also active in protesting against the war in Algeria and the increasingly frightening political response to it in France, attracted open hostility in the streets of Paris. Beauvoir felt again that she "was living in an occupied city."[22]

It is against this background that Beauvoir wrote the first volumes of her autobiography, *Memoirs of a Dutiful Daughter* and *The Prime of Life*, each of which became a best-seller immediately on publication in 1958 and 1960. These volumes, which are famously concerned with Beauvoir's escape from the condition of oppression which was the intended destiny for bourgeois women, which detail her construction of a relationship with a man founded on principles other than dominance and submission, which give the history of her and her associates' resistance to the German occupation of France, are deeply affected by Beauvoir's reaction to the Algerian war, which overtly provides the organizing structure of the third volume of the autobiography, *Force of Circumstance*, published in 1963, one year after the end of the conflict. The autobiographies are, among other things, texts that give an account of one woman's refusal of colonization. The deep strain of anti-authoritarianism, detailed in the most minute of personal actions, that runs through the volumes is partly determined by Beauvoir's response to the hated human proclivities that so disgusted her in the French involvement in Algeria.

Beauvoir's reaction to the war was shared with Sartre. The couple worked together on *Les Temps modernes* to put their journal in the service of the Algerian rebels. In *Adieux*, Sartre explained that his hatred of colonialism had begun early. By the time he was 16, he said, he looked "upon colonialism as an antihuman brutality, as an action that destroyed men for the sake of material interest."[23] Like Beauvoir, Sartre's opposition to the war was intense, and this shared passion took the couple's collaborative efforts as journalists into new dimensions of commitment. In 1960 the entire staff of *Les Temps modernes* signed the *Manifesto of the 121*, which advocated draft resistance for the French who were called upon to fight in Algeria (by 1957 there were 400,000 French troops in the colony), and assistance for the FLN. The manifesto appeared while Sartre and Beauvoir were in Brazil. There they received a letter from Claude Lanzmann (the future producer of *Shoah*, who was Beauvoir's partner from 1952 until 1958), telling them that the issue of the journal containing the manifesto had been seized and the offices of *Les Temps modernes* ransacked. There were threats of indictment against the signatories, who were the subject of violent nationalist rage (5,000 French veterans marched down the Champs Elysées chanting "Kill Sartre"). Further, the manifesto's signatories were banned from media interviews.[24] Conditions were so dangerous that Lanzmann advised the couple not to return to France, where they would be under threat both of attack and arrest. Sartre and Beauvoir decided to return to Paris in spite of the warning. They intended to turn these repressive events to their advantage and to court arrest for the sake of the attention it would gain for their position. In even this they were thwarted. Though they were charged, their case was postponed repeatedly. Mysteriously, it later simply evaporated: "'You don't arrest Voltaire,'" said de Gaulle, who was unwilling to provide the pair with the publicity they tried to secure.[25]

When the end of the conflict was near, the violence of the ultra-nationalist French OAS (Organisation de l'Armée Secrète) increased. Paris became a terrorist zone of beatings, murders, and bombings, and Beauvoir and Sartre were obvious targets. In 1961 Beauvoir moved out of the flat on the rue Schoelcher which she had bought in 1955 with the proceeds of the Prix Goncourt she had won for *The Mandarins*, and Sartre moved his mother out of the flat on the rue Bonaparte. Sartre's flat was indeed bombed, as was the building in which he and Beauvoir took refuge. Beauvoir's flat was defended by groups of students who stood vigil there in her absence. When peace came in 1962, the couple were exhausted. Sartre had been very ill. Camus, Boris Vian, Merleau-Ponty, and Richard Wright all died during the years of the war. Frantz Fanon, the Martiniquen doctor who became a key figure in the Algerian conflict, for whose classic study of colonial liberation, *The Wretched of the Earth*, Sartre had written an impassioned introduction, was also dead. Lanzmann, Beauvoir's last serious male lover, ended their relationship in 1958. With all this in mind, there seems no mystery about Beauvoir ending the third volume of her autobiography in 1963 with a bleak statement about the nonfulfillment of her hopes as a young girl: "I was gypped."[26]

Critique of Dialectical Reason

If Beauvoir both took refuge from the Algerian war and responded to it in the composition of her autobiography, Sartre used his writing of the *Critique of Dialectical Reason* in many of the same ways. In 1958 he worked on the book with a maniacal, drug-supported frenzy. As Beauvoir explained:

> It was not a case of writing as he usually did, pausing to think and make corrections, tearing up a page, starting again; for long hours at a stretch he raced across sheet after sheet without re-reading them, as though absorbed by ideas that his pen, even at that speed, couldn't keep up with; to maintain this pace, I could hear him crunching corydrane tablets, of which he managed to get through a tube a day. At the end of the afternoon he would be exhausted . . .[27]

Sartre reverted to the semiautomatic and unconsidered writing methods of his childhood for the book that would be his first major philosophical production since *Being and Nothingness*. When the latter had been completed, said Sartre, he was "fresh out of ideas."[28] Beauvoir's thought in the late 1930s had taken him as far as he could go. By the 1950s she had been working on the question of the Social Other for more than a decade, and Sartre was again ready to follow her lead, taking as his subject the relationship between individual freedom and social structures. The result, admitted Sartre, was "not a masterpiece of planning, composition, and clarity," and the cocktail of drugs he took to keep himself working at utmost speed afforded him a kind of "depraved pleasure" as he courted the possibility of overdose and death through their abuse.[29] Sartre never finished the *Critique*: it joined the long list of his uncompleted projects which included his long-promised *Ethics*, a book on Mallarmé, one on Tintoretto, a study of Nietzsche, and the final volume of the vast work on Flaubert which occupied him from 1954 until the final months of his life.[30]

Sartre had begun work on the *Critique* as early as 1953[31] at the same time as he was writing *The Words*, and laying the foundations for his Flaubert project. All three works are studies of rebellion and of the possibility of overturning the extant social order. His intention in the *Critique*, said Sartre in the introduction to the text, was to "raise one question, and only one: do we now possess the materials for constituting a structural, historical anthropology?"[32] And while Sartre was positioning himself in terms of the Marxist thought he now viewed as being of inescapable importance for the modern period, he also wished to resist the growing influence of the kind of structuralist anthropology proposed by Lévi-Strauss, who, indeed, saw Sartre as a direct rival, and who attacked the inadequacies of his analysis.[33] Sartre prioritized history and praxis as the bedrock for the formation of a group consciousness that could move forward toward commitment. Sartre's was a desperate search to find an explanation of the ways that the individual will to freedom might merge with that of the collectivity.[34] Taking Beauvoir's idea of reciprocity as his way out of the problem of the atomization of individuals who were likely to be seen simply as expressions of structural features by both Marxists and structuralists, Sartre struggled to find a way to argue for revolution that left the notion of commitment possible. In retrospect, although he agreed with Beauvoir that the *Critique* went further in its analysis than his earlier books, he also admitted that it was "very hard for me really to think of the *Critique of Dialectical Reason* as superior to *Being and Nothingness*."[35]

Sartre's unease about the volume may have been attendant not only on the fact that he reached an impasse in writing the second volume, which led to his abandoning the project, but on the lukewarm reception that the first volume received when it was published in 1960. It was noticed almost exclusively by his enemies in the Communist Party (with which he had had a *rapprochement* in the early 1950s, but which he had denounced after the suppression of the Hungarian uprising in 1956), and by the French right wing, which had always hated Sartre. If Sartre had meant to provide a philosophical background for the era of decolonization in much the same way as he had provided a philosophical alphabet of commitment and hope for the survivors of the Second World War, he was only to be greatly disappointed. The book, like *Being and Nothingness* and *Nausea*, was dedicated to Beauvoir,[36] and it is clear from her comments that she was present or nearby as Sartre actually wrote the text. Her input into this project is likely to have been significant. How much she shared Sartre's disappointment in the reception the *Critique* received is a matter for conjecture.

1968

If, by the early 1960s, French intellectual fashions had moved on to leave Sartre beached in ways that he did not fully comprehend, he was, with Beauvoir, nevertheless still an icon that the French left, and French leftist students, in particular, regarded him with enormous personal affection. In 1960, the same year that the *Critique* was greeted with such indifference, Sartre was welcomed with a mass display of warmth by the students to whom he lectured at the Sorbonne in March.[37] With their characteristic generosity and grace, both Beauvoir and Sartre were willing to be used as beloved figureheads of rebellion during the events of 1968 with which they were not deeply involved. "Like everyone else in France," said Sartre, he and Beauvoir "were caught unawares by the events of May '68."[38] To some extent, it is a measure of the degree to which the couple had retreated into their own coterie and into the demands of their ambassadorial roles, that they should have been so taken by surprise by the upheavals that would not only galvanize France but create a watchword for the possibility of protest for a generation. In another sense, their surprise was not difficult to understand: even the organizers of the events of 1968 in France seem to have been surprised by the degree of their success.

By the time of the May demonstrations which involved students, factory workers, and various leftist political factions, Sartre was 62 years old and Beauvoir was 60. Neither could have anticipated fully the directions in which the events of May would pull them. Sartre's alignment with the Maoists and Beauvoir's important association with a number of feminist causes grew directly out of their reactions to the demonstrations to which they bore important witness, but of which they were not an integral part.

Their intellectual passion, along with their customary cult of youth, put them in a good position to respond energetically to the new circumstances at an age when many would be thinking of anything rather than deepening commitment to new political formations.[39]

The initial focus for the protests of May 1968 were centered around causes that Beauvoir and Sartre fully supported. Since 1966, the couple had attended tribunals and conferences called to protest against American action in Vietnam. Beauvoir and Sartre also were interested in the educational reforms that the student protestors demanded, and *Les Temps modernes* had covered the issues sympathetically. But while the pair was sympathetic to the international and local aims of the students, they also felt removed from them. Similarly, while they were quick to sign protests denouncing police brutality against the students and to support manifestos that called for further action on their behalf, the couple felt themselves useful only on the fringes of the action. Again, as was the case during the Algerian war, they did what they could. Sartre gave an interview on Radio Luxemburg on May 12, six days after the start of the demonstrations, in which he spoke in support of student use of violence.[40] By May 13, half a million demonstrators surged through the Paris streets, Beauvoir among them.[41] On May 20, Sartre spoke at the Sorbonne but felt that, though he was willing to support the students, he was, in some sense, irrelevant to their needs. "In fact," he said, "I had nothing to say to them, not having been a student for a great while and not being a teacher. There was nothing that qualified me to speak."[42] What he did find to say was unexpected and he was cheered less when he left the rostrum than when he ascended it. But he was now linked with the extraordinary youth movement in interesting ways. For the first time since the failure of the RDR he now felt he was politically active.[43] Sartre's later willingness to serve as an editor of the banned Maoist paper, *La Cause du peuple*, and his general involvement with Maoism through his last secretary, Pierre Victor (Benny Lévy), were seen by many as evidence of his fading powers of judgment, and as the willingness of an old man to be led down ridiculous paths. Sartre saw things otherwise. His intimate contact with and interest in the ideas of the young Maoists who emerged from the events of 1968 kept him intellectually alive and fed the craving for youth that had always been a significant part of his character. In a touching gesture of solidarity with the ideals of the student liberationists of 1968, Sartre never again wore a suit or a tie.[44] It was widely acknowledged that Sartre's existentialist ideas established him as one of the intellectual godfathers of the events he so applauded, and like a good godfather, he took a proud interest in his youthful charges.

Beauvoir's reaction to 1968 was longer in the making but no less profound. Although she signed the various petitions and manifestos of

1968 along with Sartre and contributed various articles and statements to the press in support of the student demonstrators, just as she joined Sartre in his support for *La Cause du peuple* and courted arrest with him for distributing copies of the newspaper in the street (as well as accepting the directorship of another banned leftist paper, *L'Idiot internationale*), the most lasting impact of 1968 on Beauvoir was in her increasing affiliation with the feminist movements that grew out of the events in France, and out of the civil rights and anti-war movements in the United States. Although Beauvoir had been active in occasional ways in the campaign for abortion rights since 1960, and despite her passionate defense of Djamila Boupacha during the Algerian conflict, she was not ready to retract her famous personal disclaimer of the feminist label made in *The Second Sex* until 1972, a year after she signed the MLF's (Mouvement de la Libération des Femmes) *Manifesto of the 343*, a document published in *Le Nouvel Observateur*, in which 343 women publicly announced that they had had illegal abortions.[45] Beauvoir later denied that she had, in fact, ever terminated a pregnancy, but, whatever the truth of the matter, the important point was that the signing of *The Manifesto of the 343* marked Beauvoir as committed to feminist politics in ways that would have been unlikely earlier in her life.[46] Like Sartre, and with the same generosity, Beauvoir committed herself to the struggles of the young in ways that profoundly affected her own life, ways that, again like Sartre, brought her into contact with the young women who would treat her in her old age as a friend, an exemplar, and a force to be considered even when rejected.[47]

The events of 1968 brought Sartre and Beauvoir back into the mainstream of alternative practice and politics in France. Directly and indirectly it linked them with the young who helped them continue their cult of youth, their commitment, and their passion for freedom into their last years.

Adieux

Sartre died on April 15, 1980, after a protracted and harrowing decline which Beauvoir, with her characteristic honesty, chronicled in detail in *Adieux*, her book of brutally loving farewell to the companion of her adult years. Fifty thousand mourners accompanied Sartre's coffin to its grave in the Montparnasse Cemetery on April 19. Beauvoir was so overcome by the whisky and Valium she had taken to help her face the occasion that she could scarcely stand, and had to be physically supported to a chair by Sartre's grave, where she sat, stupefied, for ten minutes while the crowd seethed around her. She was too ill to attend Sartre's cremation on April 23, or the final interment of his ashes at the Montparnasse Cemetery on the same day. In her grief and in the exhaustion that followed her long vigil over the dying Sartre, Beauvoir fell seriously ill with pneumonia.[48] During the month of hospitalization which followed, Beauvoir's sorrow and depression were extreme. Her own health was bad and her mental state no better as she brooded on the way she felt that Sartre had been stolen from her in his last years by his adopted daughter, Arlette Elkaïm, and his Maoist secretary, Benny Lévy. Squabbles with Arlette over Sartre's literary and personal effects, which began days after his death and were to continue for the rest of Beauvoir's life, made for additional bitterness. Her doctors were pessimistic about Beauvoir's chances of a full recovery, but, with the help of the young woman whom she decided to adopt as her daughter and literary executor while in hospital, Sylvie Le Bon, Beauvoir slowly pulled herself out of her misery to take up her life again, and to write her final memorial to Sartre and to their relationship, *Adieux*. The opening dedication of the book deserves to be cited at length, as Beauvoir takes a moving leave of the man who had made her life possible, just as she had made his:

> This is the first of my books – the only one no doubt – that you will not have read before it is printed. It is wholly and entirely devoted to you; and you are not affected by it.
> When we were young and one of us gained a brilliant victory over the other in an impassioned argument, the winner used to say, "There you are in your little box!" You are in your little box; you will not come out of it and I shall not join you there. Even if I am buried next to you there will be no communication between your ashes and mine.[49]

With her bitter realism, Beauvoir wrote her long goodbye to her lover as the final instalment of her own mythmaking with regard to their relationship, as well as a characteristic writer's exercise in dealing with her own grief. When *Adieux* appeared in 1981, it was savaged by the reviewers for its detailed account of Sartre's physical decline.[50] It is, in fact, a profoundly moving document that faces life and death with equally clear-eyed vision.

Beauvoir died on April 15, 1986. A crowd of 5,000 mourners followed her coffin to the grave she shares with Sartre in the Montparnasse Cemetery.[51]

Whatever accounts Sartre, and especially Beauvoir, gave of their intellectual partnership during their lives, it is clear that Beauvoir, in particular, was careful to leave behind after her death copious documentary material in the form of letters and journals which would invite quite other readings of that relationship than the one to which she devoted so much energy, so many printed words, and so many hours of interviews while she still lived. "An experience," said Beauvoir in *Force of Circumstance*, musing on the difference between life and accounts of life, "is not a series of facts."[52] She was obviously, if only partly, right. The newly known facts which stand behind this reinterpretation of the balance of Sartre and Beauvoir's mutual philosophical and personal indebtedness place previous accounts of the couple's intellectual and intimate experience in a radically new light. And we remain convinced that this revisionary biography of one of history's most extraordinary intellectual partnerships will stand as an important step in the remaking of a twentieth-century legend.

NOTES

Chapter 1: The Precocious Plagiarist

1. See John Gerassi, *Jean-Paul Sartre: Hated conscience of his century*, vol. 1, *Protestant or Protester*, University of Chicago Press: London, 1989, p. 90, and Deirdre Bair, *Simone de Beauvoir: A biography*, Jonathan Cape: London, 1990, p. 628.
2. Gerassi, *Jean-Paul Sartre*, p. 64.
3. Jean-Paul Sartre, *The Words*, trans. Bernard Frechtman, George Braziller: New York, 1964, p. 22.
4. Ibid., pp. 11–12.
5. Ibid., pp. 21, 22, 26, 27.
6. Ibid., pp. 23, 31, 34–5.
7. Ibid., p. 35.
8. Ibid., pp. 41, 43.
9. Ibid., pp. 50–1.
10. Ibid., pp. 59, 61, 64, 70.
11. Ibid., p. 118–25.
12. Ibid., pp. 103, 104.
13. Ibid., p. 104.
14. Ibid., p. 81.
15. Ibid., p. 83.
16. Ibid., p. 86.
17. Ibid., pp. 87, 132.
18. Jean-Paul Sartre, *War Diaries: Notebooks from a phoney war, November 1939–March 1940*, trans. Quintin Hoare, Verso: London, 1984, p. 70.
19. *The Words*, pp. 144–5.
20. Ibid., pp. 145, 153.
21. Ibid., p. 155.

22. Ibid., pp. 156, 158.
23. Ibid., pp. 166, 167.
24. Ibid., p. 167.
25. Ibid., p. 170.
26. Ibid., pp. 181–2, 215.
27. Ibid., pp. 217, 218 (emphasis added).
28. *War Diaries*, p. 273.
29. *The Words*, p. 222.
30. Michel Contat and Michel Rybalka (eds), *The Writings of Jean-Paul Sartre*, vol. 1, *A Bibliographical Life*, trans. Richard C. McCleary, Northwestern University Press: Evanston, 1974, p. 4.
31. Jean-Paul Sartre, *Sartre by Himself*, trans. Richard Seaver, Urizen: New York, 1978, p. 8. Gerassi, *Jean Paul Sartre*, p. 57.
32. *Sartre by Himself*, p. 8.
33. Simone de Beauvoir, *Adieux: A farewell to Sartre*, trans. Patrick O'Brian, Penguin: Harmondsworth, 1985, pp. 143, 144 (emphasis added).
34. *War Diaries*, p. 268.
35. *Adieux*, p. 242.
36. Gerassi, *Jean-Paul Sartre*, pp. 60–1.
37. *Adieux*, p. 146.
38. Gerassi, *Jean-Paul Sartre*, p. 62.
39. *Adieux*, p. 147. *Sartre by Himself*, p. 11.
40. *War Diaries*, p. 279.
41. Ibid., p. 268.
42. *Adieux*, p. 350.
43. Jean-Paul Sartre, "Self-portrait at seventy," *Life/Situations: Essays written and spoken*, trans. Paul Auster and Lydia Davis, Pantheon: New York, 1977, p. 37. *Adieux*, p. 219.
44. "Self-portrait at seventy," *Life/Situations*, p. 37.
45. *Adieux*, p. 190.
46. Ibid., pp. 132, 133, 431.
47. Ibid., p. 242.
48. Annie Cohen-Solal, *Sartre: A life*, Heinemann: London, 1987, p. 51.
49. *Adieux*, p. 149.
50. Ibid., p. 133.
51. Gerassi, *Jean-Paul Sartre*, p. 64. *Adieux*, p. 132.
52. *Adieux*, p. 138.
53. Gerassi, *Jean-Paul Sartre*, p. 66.
54. From *Situations*, cited in Ronald Hayman, *Writing Against: A biography of Sartre*, Weidenfeld and Nicolson: London, 1986, p. 41.
55. *War Diaries*, p. 73.
56. *Adieux*, p. 143. *War Diaries*, pp. 80, 74, 80.
57. Gerassi, *Jean-Paul Sartre*, p. 71.
58. Ibid., p. 90, and Bair, *Simone de Beauvoir*, p. 628.
59. Hayman, *Writing Against*, p. 44.
60. From *Situations IV*, cited in Gerassi, *Jean-Paul Sartre*, pp. 66–7.

61. *War Diaries*, p. 85.
62. *Adieux*, pp. 149, 245, 256.
63. Simone de Beauvoir, *The Prime of Life*, trans. Peter Green, Penguin: Harmondsworth, 1965, pp. 66, 67.
64. Gerassi, *Jean-Paul Sartre*, p. 79.
65. Hayman, *Writing Against*, p. 49.
66. Ibid., p. 58. Gerassi, *Jean-Paul Sartre*, p. 76.
67. Cohen-Solal, *Sartre*, p. 73.

Chapter 2: The Phenomenon who could not be Judged

1. Simone de Beauvoir, *Memoirs of a Dutiful Daughter*, trans. James Kirkup, Penguin: Harmondsworth, 1963, p. 5.
2. For an account of feminist suspicions regarding Beauvoir on these grounds see Toril Moi, *Feminist Theory and Simone de Beauvoir*, Blackwell: London, 1990.
3. For information on Beauvoir's life which does not appear in her own autobiographies, we are particularly indebted to Bair, *Simone de Beauvoir*; to Claude Francis and Fernande Gontier, *Simone de Beauvoir*, trans. Lisa Nesselson, Mandarin: London, 1989; and to Margaret Crosland, *Simone de Beauvoir: The woman and her work*, Heinemann: London, 1992.
4. *Memoirs*, p. 11.
5. Ibid., p. 9.
6. Ibid., pp. 11, 12, 14.
7. Bair, *Simone de Beauvoir*, p. 27. See also pp. 22–8 for information on Beauvoir's parents.
8. Francis and Gontier, *Simone de Beauvoir*, p. 371, from an interview on September 20, 1985.
9. *Memoirs*, pp. 35–6.
10. Ibid., pp. 36, 37, 38.
11. Bair, *Simone de Beauvoir*, p. 28.
12. *Memoirs*, pp. 59, 296.
13. Ibid., pp. 53, 69.
14. Ibid., p. 52. Francis and Gontier, *Simone de Beauvoir*, p. 25.
15. *Memoirs*, pp. 36, 37.
16. As Hélène de Beauvoir said in an interview: "Our mother was the hen who had given birth to ducks . . . She was very proud of her daughters; she wanted to live through us, but she also wanted us to live for her, and we wouldn't do that." Bair, *Simone de Beauvoir*, p. 108.
17. See ibid., pp. 42–3, for information on Beauvoir's education.
18. Ibid., pp. 43, 625.
19. *Memoirs*, p. 122.
20. Ibid., pp. 121–2.
21. Bair, *Simone de Beauvoir*, p. 41. *Memoirs*, pp. 44–5, 56.
22. *Memoirs*, pp. 90–1.
23. Ibid., p. 140. Elaine Showalter notes the importance of *The Mill on the Floss* for Beauvoir and suggests that *Memoirs of a Dutiful Daughter* is "a structured

homage to Eliot's novel" in *Sexual Anarchy: Gender and culture at the fin de siècle*, Bloomsbury: London, 1991, pp. 70–1.

24. Information on the war years taken from Bair, *Simone de Beauvoir*, pp. 46–7; Francis and Gontier, *Simone de Beauvoir*, pp. 24, 30–1; and *Memoirs*, pp. 62–5.

25. *Memoirs*, p. 64.

26. Beauvoir's concern with death has long been recognized as central to her work. See, for example, Elaine Marks's pioneering study, *Simone de Beauvoir: Encounters with death*, Rutgers University Press: New Brunswick, NJ, 1973.

27. *Memoirs*, pp. 137, 138.

28. From Francis Jeanson, *Simone de Beauvoir ou l'entreprise de vivre*, cited in Bair, *Simone de Beauvoir*, p. 59.

29. *Memoirs*, pp. 41, 107, 108, 111.

30. Ibid., p. 104. That this lack of a dowry might mean anything at all to Beauvoir (the dowryless daughter of a dowryless mother) is fascinating. Indeed, it is startling to see Beauvoir, in an interview late in her life, giving her lack of a dowry as the reason for not marrying Sartre, instead of the highly principled accounts of her reasons for her distrust of marriage that she gives elsewhere. That such a reactionary institution might have played any part in the association of Beauvoir and Sartre is almost unthinkable in the light of the relational values of the late twentieth century which they did much to help form. (See also Bair, *Simone de Beauvoir*, p. 156.)

31. See Alice Schwarzer, *Simone de Beauvoir Today: Conversations 1972–1982*, trans. Marianne Howarth, Chatto & Windus: London, 1984, and Hélène Vivienne Wenzel (ed.), "Simone de Beauvoir: Witness to a century", *Yale French Studies*, no. 72, 1986, pp. 5–32.

32. Simone de Beauvoir, *The Ethics of Ambiguity*, trans. Bernard Frechtman, Citadel: New York, 1948, p. 35.

33. Ibid., p. 39.

34. Francis and Gontier, *Simone de Beauvoir*, p. 41.

35. *Memoirs*, p. 92.

36. Ibid., p. 93.

37. Ibid., pp. 112, 113, 114, 119.

38. Ibid., pp. 255–6.

39. Bair, *Simone de Beauvoir*, p. 152.

40. *Memoirs*, p. 60.

41. Hélène de Beauvoir recalled that Jacques

    was the only young man we saw in such an informal manner . . . We saw no young men during those years except our cousins, and Jacques was the only one to come to our house in the evenings when we were alone as a family, to sit with us and talk as if we were his equals. He had an aura of culture and independence which was very exciting, and we listened attentively to everything he said. (Bair, *Simone de Beauvoir*, p. 84)

42. *Memoirs*, pp. 158, 173.

43. Ibid., p. 202, and Bair, *Simone de Beauvoir*, p. 102.

44. *Memoirs*, p. 201.

45. Ibid., pp. 218, 315–16, 346–7.
46. Ibid., p. 348.
47. Ibid., p. 104.
48. Ibid., pp. 159–60; Bair, *Simone de Beauvoir*, p. 92.
49. *Memoirs*, pp. 158, 160.
50. Ibid., pp. 177, 179.
51. Bair, *Simone de Beauvoir*, p. 94.
52. Francis and Gontier, *Simone de Beauvoir*, p. 51.
53. See Michèle Le Doeuff's perceptive analysis of the importance of the imitation of key teachers in academic growth, and the special dangers this holds for women, in her article "Women and philosophy," trans. Debby Pope, *Radical Philosophy*, no. 17, 1977, pp. 2–11.
54. *Memoirs*, pp. 207, 233, 234.
55. Ibid., p. 208.
56. Ibid., pp. 195, 211.
57. Ibid., pp. 278, 284–5, 289.
58. Francis and Gontier, *Simone de Beauvoir*, p. 79.
59. *Memoirs*, pp. 234–5.
60. Ibid., p. 239. Beauvoir's later reaction was quite different. In 1983 she was embarrassed by her youthful attitudes:

    > Simone Weil was right to dismiss me like that. It took me many years to free myself from what I called in my memoirs 'the bonds of my class'. I know that even today there are many who accuse me of behaviour instilled by 'the bonds of class', especially some feminist women. Perhaps they are right, and one never overcomes the class into which one is born. I don't know. (Bair, *Simone de Beauvoir*, p. 120)

61. Ibid., p. 125.
62. *Memoirs*, pp. 309–10.
63. Ibid., pp. 310–14.
64. Ibid., pp. 313, 324–5.
65. Ibid., p. 323.
66. Francis and Gontier, *Simone de Beauvoir*, p. 82.
67. *Memoirs*, p. 295.

Chapter 3: The Oath

1. See Gerassi, *Jean-Paul Sartre*, p. 90, and Beauvoir, *Memoirs*, pp. 310, 319.
2. *Memoirs*, p. 321. Hayman, *Writing Against*, p. 70.
3. *Memoirs*, pp. 331–2.
4. Ibid., p. 331.
5. Gerassi, *Jean-Paul Sartre*, p. 91, and Bair, *Simone de Beauvoir*, pp. 144, 142–3.
6. *Memoirs*, pp. 322–4.
7. It must be noted that Sartre's fear of ending his life as a student seems a more likely cause of his examination failure in the previous year than an excess of originality.

8. *Memoirs*, pp. 310, 334.
9. Sartre, *Sartre by Himself*, pp. 21–2.
10. *Memoirs*, pp. 334–5.
11. *Sartre by Himself*, p. 23.
12. *Memoirs*, p. 337.
13. The remark is misused in a variety of ways. For example, Michèle Le Doeuff states that Sartre's remark is "the conclusion of the first volume of Simone de Beauvoir's autobiography" ("Women and philosophy," *Radical Philosophy*, p. 8). In fact, the quotation from Sartre appears more than 10,000 words from the end of Beauvoir's book, and Sartre figures neither in the conclusion nor the penultimate section of the volume.
14. *Memoirs*, p. 339.
15. Josée Dayan and Malka Ribowska, *Simone de Beauvoir*, Gallimard: Paris, 1979, p. 20.
16. Gerassi, *Jean-Paul Sartre*, p. 90, from an interview with John Gerassi, March 26, 1971. Maheu, in another interview with Gerassi, confirmed that he had been Beauvoir's first lover (see Bair, *Simone de Beauvoir*, p. 628).
17. *Le Nouvel Observateur*, March 21, 1976, p. 15, cited in Gerassi, *Jean-Paul Sartre*, p. 91.
18. *Memoirs*, p. 342.
19. Ibid., pp. 319, 324–6.
20. Ibid., pp. 339–41, 340.
21. *War Diaries*, p. 75. Sartre as a suitor was disinclined to understate his accomplishments, past and future. In 1926 he wrote to Simone Jollivet that he had finished his first novel when he was 8, had created several philosophical systems before he was 17, and had also composed symphonies. See *Lettres au Castor et à quelques autres*, vol. 1, *1926–1939*, ed. Simone de Beauvoir, Gallimard: Paris, 1983, p. 9.
22. See Gerassi, *Jean-Paul Sartre*, p. 91; Bair, *Simone de Beauvoir*, pp. 145–6; and Francis and Gontier, *Simone de Beauvoir*, p. 92.
23. Bair, *Simone de Beauvoir*, pp. 155–6.
24. *Memoirs*, p. 345.
25. Francis and Gontier, *Simone de Beauvoir*, p. 92.
26. *Prime of Life*, pp. 13–14.
27. Ibid., p. 14, and Bair, *Simone de Beauvoir*, pp. 149–50.
28. Bair, *Simone de Beauvoir*, pp. 150, 155, 156.
29. *War Diaries*, p. 75.
30. *Memoirs*, pp. 330, 346, 347, 355, 358.
31. Ibid., pp. 358–9, 360.
32. *Prime of Life*, p. 9. The circumstances involved in the hostility to Zaza's marriage included the uncovering of Merleau-Ponty's illegitimacy, of which he, himself, had hitherto been unaware.
33. *Memoirs*, pp. 314, 326 (emphasis added).
34. *Prime of Life*, p. 12.
35. *Memoirs*, pp. 144, 145, 345.
36. Ibid., pp. 340–1.

37. Ibid., pp. 342–3.
38. *War Diaries*, p. 282. *Memoirs*, p. 343.
39. *Memoirs*, pp. 343, 344.
40. Mary Evans, *Simone de Beauvoir: A feminist mandarin*, Tavistock: London, 1985, p. 16.
41. *Prime of Life*, p. 24.
42. Ibid., p. 22.
43. Ibid., p. 23.
44. Ibid., pp. 23–4.
45. *Letters to Sartre*, trans. and ed. Quintin Hoare, Radius: London, 1991, pp. 4–5.
46. *Prime of Life*, p. 56.
47. Ibid., p. 8.

Chapter 4: Breakdowns and Beginnings

1. Jean-Paul Sartre, *Witness to my Life: The letters of Jean-Paul Sartre to Simone de Beauvoir, 1926–1939*, ed. Simone de Beauvoir, trans. Lee Fahnestock and Norman MacAfee, Scribner's: New York, 1992, p. 33.
2. Jean-Paul Sartre, *Being and Nothingness: An essay on phenomenological ontology*, trans. Hazel E. Barnes, Philosophical Library: New York, 1956, p. 39.
3. *Prime of Life*, p. 59.
4. Ibid., pp. 53–4; Francis and Gontier, *Simone de Beauvoir*, pp. 113–14.
5. *Prime of Life*, p. 54.
6. Ibid., pp. 34–8, 55.
7. Ibid., pp. 44–6.
8. Ibid., pp. 62, 63, 64.
9. Ibid., p. 56.
10. Ibid., pp. 60, 61, 62.
11. *Witness to my Life*, pp. 35–41.
12. *Memoirs*, p. 329.
13. *Witness to my Life*, p. 41.
14. *Prime of Life*, pp. 88, 89, 90, 92–3.
15. Ibid., pp. 93, 100.
16. Ibid., p. 103.
17. Ibid., p. 106.
18. Ibid., pp. 124, 125.
19. Ibid., p. 128.
20. Gerassi, *Jean-Paul Sartre*, p. 113.
21. *Prime of Life*, pp. 135–6.
22. Ibid., pp. 158, 183–4.
23. Ibid., p. 201.
24. Sartre, *Sartre by Himself*, pp. 29–30.
25. *Prime of Life*, pp. 220, 201.
26. Jean-Paul Sartre, *The Transcendence of the Ego: An existentialist theory of consciousness*, trans. Forrest Williams and Robert Kirkpatrick, Noonday Press: New York, 1957, p. 31.

27. *Prime of Life*, p. 40.
28. Ibid., pp. 106, 138.
29. Ernest Hemingway, *A Farewell to Arms*, Scribner's: New York, 1929, p. 185.
30. Hemingway, *A Farewell to Arms*, p. 227.
31. Jean-Paul Sartre, *Nausea*, trans. Robert Baldick, Penguin: Harmondsworth, 1965, p. 252.
32. *Prime of Life*, p. 284.
33. Ibid., pp. 201, 207.
34. Ibid., p. 212.
35. Ibid., pp. 209, 210–11, 212.
36. Ibid., pp. 217–20.
37. Schwarzer, *Simone de Beauvoir Today*, p. 58.
38. *Sartre by Himself*, p. 28.
39. *Prime of Life*, pp. 245, 246.
40. Ibid., pp. 240, 242.
41. *Adieux*, p. 306.
42. See the remarks on the relationship throughout *Letters to Sartre*. See also *Prime of Life*, pp. 257–63.
43. *Prime of Life*, p. 260.
44. Ibid., pp. 259–61.
45. *Adieux*, p. 172.
46. *Prime of Life*, pp. 275–85.
47. *Adieux*, p. 159.
48. Bair, *Simone de Beauvoir*, p. 202.
49. *Prime of Life*, p. 293.
50. *Witness to my Life*, pp. 92–5.
51. *Prime of Life*, pp. 293–302.
52. Hayman, *Writing Against*, p. 127, and Cohen-Solal, *Sartre*, p. 121.
53. *Prime of Life*, p. 241.
54. Bair, *Simone de Beauvoir*, p. 209.
55. Ibid., p. 208.
56. *Prime of Life*, pp. 225, 319, 365.
57. Ibid., pp. 327, 336, 337, 346.
58. Bair, *Simone de Beauvoir*, p. 215.
59. See for example Schwarzer, *Simone de Beauvoir Today*, pp. 112–13.
60. *Letters to Sartre*, p. 260.
61. *Prime of Life*, p. 25.
62. Ibid., p. 337.
63. *Lettres au Castor*, vol. 1, p. 274.

Chapter 5: The True Philosopher and the Man with the Black Gloves

1. Margaret A. Simons, "Beauvoir and Sartre: The philosophical relationship," *Yale French Studies*, no. 72, 1986, pp. 165–79.
2. Ibid., p. 168.

3. Simone de Beauvoir, *She Came to Stay*, trans. Yvonne Moyse and Roger Senhouse, Flamingo: London, 1984.

4. Ibid., pp. 52, 54.

5. *Being and Nothingness*, p. 304.

6. *Letters to Sartre*, p. 200.

7. Ibid., p. 158.

8. Simone de Beauvoir, *Journal de guerre: septembre 1939–janvier 1941*, ed. Sylvie Le Bon de Beauvoir, Gallimard: Paris, 1990, pp. 270–3, 280.

9. Hazel Barnes's reading of *She Came to Stay* in 1959 is a partial and interesting exception. As the English translator of *Being and Nothingness*, Barnes was particularly well placed to identify the philosophical content in Beauvoir's novel and succeeded admirably in delineating its theory of Others. But all the rest – and, as shall be seen, it is a great deal – she missed. Barnes, however, was working with the severe handicap of not knowing the order in which the two books had been written. Although she attributes all the ideas and images common to the two works to Sartre, a striking characteristic of her reading is an acute unease that she may be mistaken on this very point. See Barnes, *The Literature of Possibility: A study in humanistic existentialism*, Tavistock: London, 1959. For other kinds of readings of Beauvoir's fiction see, especially, Elizabeth Fallaize, *The Novels of Simone de Beauvoir*, Routledge: London, 1988, and Anne Whitmarsh, *Simone de Beauvoir and the Limits of Commitment*, Cambridge University Press: Cambridge, 1981.

10. *She Came to Stay*, p. 1.

11. Ibid., p. 1.

12. Ibid., pp. 1–2.

13. Ibid., p. 2.

14. *Prime of Life*, p. 13.

15. Bertrand Russell, *The Problems of Philosophy*, Oxford University Press: London, 1967, p. 6.

16. *She Came to Stay*, pp. 2–3.

17. *Transcendence of the Ego*, p. 103.

18. Jean-Paul Sartre, *The Emotions: Outline of a theory*, trans. Bernard Frechtman, Philosophical Library: New York, 1948, pp. 11, 18.

19. *Transcendence of the Ego*, p. 105.

20. *She Came to Stay*, pp. 6–7.

21. Arthur C. Danto, *Sartre*, Fontana: London, 1991, p. 1.

22. *Prime of Life*, p. 328.

23. *Letters to Sartre*, pp. 20–1.

24. Ibid., p. 178.

25. Ibid., pp. 37, 46.

26. See Sartre's *War Diaries*, pp. 160–2, for a long passage from a letter Sartre received from Bost.

27. *Letters to Sartre*, pp. 54, 86, 92, 120, 149–50.

28. *Witness to my Life*, p. 230.

29. *War Diaries*, pp. 170–1.

30. Hayman, *Writing Against*, p. 145.

31. Cohen-Solal, *Sartre*, p. 139.
32. *The Words*, pp. 146, 153.
33. *Witness to my Life*, p. 249.
34. *Lettres au Castor*, vol. 2, p. 27. Translation taken from Cohen-Solal, *Sartre*, p. 142.
35. *Journal de guerre*, pp. 270–3, 280.
36. *War Diaries*, pp. 197, 208.
37. *Being and Nothingness*, pp. 527–9.
38. *War Diaries*, pp. 208–9.
39. *She Came to Stay*, p. 51.
40. *War Diaries*, pp. 209, 214.
41. *She Came to Stay*, pp. 69, 81.
42. Ibid., p. 135.
43. *War Diaries*, pp. 229, 230.
44. Ibid., p. 11.
45. *She Came to Stay*, p. 131.
46. *War Diaries*, p. 258.
47. *Letters to Sartre*, pp. 275, 277.
48. *Adieux*, p. 304.
49. *Witness to my Life*, p. 198.
50. Hayman, *Writing Against*, p. 128.
51. *Lettres au Castor et à quelques autres*, ed. Simone de Beauvoir, vol. 2, *1940–1963* Gallimard: Paris, 1983, pp. 94, 111.
52. *Prime of Life*, p. 23.
53. Bair, *Simone de Beauvoir*, p. 207.
54. Sylvia Lawson, "All in the family," *London Review of Books*, December 3, 1992, p. 16.
55. *She Came to Stay*, pp. 7, 43–4.
56. Gerassi, *Jean-Paul Sartre*, p. 158.
57. *Prime of Life*, pp. 315–18, and Bair, *Simone de Beauvoir*, pp. 228–32.
58. Bair, *Simone de Beauvoir*, p. 228.
59. Ibid., pp. 228–9.
60. Ibid., pp. 601–2, 678–9.
61. *Letters to Sartre*, p. xi.

Chapter 6: The War Years

1. As Sartre wrote to Beauvoir on September 20, 1939, "Vous semblez comme moi ne pas 'réaliser' la guerre," *Lettres au Castor*, vol. 1, p. 305.
2. *Prime of Life*, p. 359.
3. Ibid., p. 359.
4. Ibid., p. 362.
5. Ibid., p. 369.
6. *War Diaries*, pp. 78, 88.
7. *Adieux*, p. 387.
8. *Prime of Life*, p. 437.

9. James D. Wilkinson, *The Intellectual Resistance in Europe*, Harvard University Press: London, 1981, p. 25.
10. *Prime of Life*, pp. 445, 450–1.
11. Ibid., p. 457; *Lettres au Castor*, vol. 2, pp. 282–3; *Journal de guerre*, p. 344.
12. *Prime of Life*, pp. 452, 457.
13. *Adieux*, p. 389. *Prime of Life*, p. 427.
14. *Lettres au Castor*, vol. 1, p. 307.
15. Ibid., p. 308.
16. Hayman, *Writing Against*, p. 162.
17. Francis and Gontier, *Simone de Beauvoir*, p. 190.
18. Bair, *Simone de Beauvoir*, p. 246.
19. Wilkinson, *The Intellectual Resistance in Europe*, p. 36. From an interview with Sartre, June 22, 1974.
20. *Prime of Life*, pp. 407–8, 466, 478, 479, 480.
21. Ibid., p. 482.
22. See ibid., p. 482; Cohen-Solal, *Sartre*, p. 166; Hayman, *Writing Against*, pp. 173–4; Bair, *Simone de Beauvoir*, p. 252.
23. Crosland, *Simone de Beauvoir*, p. 279.
24. Bair, *Simone de Beauvoir*, pp. 253–4.
25. *Prime of Life*, pp. 465, 484.
26. Ibid., p. 490.
27. Bair, *Simone de Beauvoir*, pp. 264, 284.
28. *Prime of Life*, p. 533.
29. Ibid., pp. 526–7.
30. Bair, *Simone de Beauvoir*, p. 272.
31. *Prime of Life*, p. 536; Wilkinson, *The Intellectual Resistance in Europe*, pp. 43–4.
32. *Prime of Life*, p. 536.
33. Ibid., p. 500.
34. Ibid., p. 539; Wilkinson, *The Intellectual Resistance in Europe*, p. 41.
35. Hayman, *Writing Against*, p. 197.
36. *Letters to Sartre*, pp. 377–8.
37. As Cohen-Solal notes in *Sartre* (p. 188), *Being and Nothingness* took years to find its readership. There was only one mention of it in an article in 1943. Two reviews appeared in 1944; nine in 1945; and over fifteen in 1946. See Contat and Rybalka, *The Writings of Jean-Paul Sartre*, vol. 1, *A Bibliographical Life*, pp. 81–4, for the publishing history of the book.
38. Hayman, *Writing Against*, p. 202.
39. *Prime of Life*, p. 558.
40. Ibid., pp. 556, 558.
41. Ibid., p. 559.
42. Bair, *Simone de Beauvoir*, p. 287.
43. *Prime of Life*, p. 548.
44. "Pyrrhus and Cynéas," trans. Christopher Freemantle, *Partisan Review*, vol. 3, part 3, 1946, p. 333.
45. *Prime of Life*, p. 549.
46. Ibid., pp. 550, 562.

47. Bair, *Simone de Beauvoir*, pp. 280, 291–2. For an excellent study of the intellectual harmonies and discords between Sartre and Camus, see Germaine Brée, *Camus and Sartre: Crisis and commitment*, Calder & Boyars: London, 1974.
48. Hayman, *Writing Against*, p. 204.
49. *Prime of Life*, p. 562.
50. *Adieux*, p. 217.
51. *Prime of Life*, p. 581.
52. Ibid., p. 583.
53. Bair, *Simone de Beauvoir*, pp. 293–4, 641.
54. *Prime of Life*, p. 597.
55. Bair, *Simone de Beauvoir*, p. 251.

Chapter 7: Existentialist Heroes

1. *Prime of Life*, p. 548.
2. Simone de Beauvoir, *Force of Circumstance*, trans. Richard Howard, Penguin: Harmondsworth, 1968, pp. 11, 12.
3. Wilkinson, *The Intellectual Resistance in Europe*, pp. 56–7.
4. Contat and Rybalka, *The Writings of Jean-Paul Sartre*, vol. 1, *A Bibliographical Life*, p. 104.
5. Jean-Paul Sartre, *Sartre on Theater*, ed. Michel Contat and Michel Rybalka, trans. Frank Jellinek, Quartet: London, 1976, pp. vii, viii.
6. Sartre, "On dramatic style," *Sartre on Theater*, p. 14.
7. See ibid., "Epic theater and dramatic theater," pp. 77–120. Lecture delivered at the Sorbonne, March 29, 1960.
8. Ibid., "Forgers of myth," p. 39. Lecture delivered in New York, 1946.
9. *Force of Circumstance*, p. 21.
10. Simone de Beauvoir, *Old Age*, trans. Patrick O'Brian, Penguin: Harmondsworth, 1977, p. 505.
11. *Prime of Life*, p. 606; *Force of Circumstance*, p. 21.
12. *Force of Circumstance*, pp. 16, 18, 40.
13. Ibid., pp. 21–2, 24.
14. Hayman, *Writing Against*, p. 218.
15. *Sartre by Himself*, pp. 62, 63, 64.
16. *Force of Circumstance*, p. 22.
17. Ibid., p. 25.
18. Ibid., p. 26.
19. *Adieux*, p. 152.
20. See Francis and Gontier, *Simone de Beauvoir*, pp. 212–13. See Hayman, *Writing Against*, pp. 220–2, for information on Sartre's American trip.
21. *Adieux*, pp. 305, 306.
22. Bair, *Simone de Beauvoir*, p. 302. Other information on Sartre's affair with Dolorès can be found in Cohen-Solal, *Sartre*, pp. 236–8, and Hayman, *Writing Against*, p. 221.
23. *Force of Circumstance*, pp. 19, 38.

24. Ibid., pp. 77–8. In *Force of Circumstance* Dolorès is referred to as "M." See also Bair, *Simone de Beauvoir*, pp. 303–5, in which Bair argues convincingly that the incident cited took place in 1945 rather than in 1946, as Beauvoir claimed in her memoirs.

25. *Force of Circumstance*, p. 46.

26. *Adieux*, pp. 247, 249, 250–1.

27. Hayman, *Writing Against*, p. 226.

28. *Force of Circumstance*, p. 46.

29. Ibid., p. 48.

30. Brée, *Camus and Sartre*, p. 31.

31. *Force of Circumstance*, p. 223.

32. See, for example, Beauvoir's explanation to Deirdre Bair of her reasons for pretence regarding her own sexual behavior (Bair, *Simone de Beauvoir*, pp. 302, 642).

33. *Force of Circumstance*, pp. 46, 54, 55.

34. Ibid., p. 24.

35. Francis and Gontier, *Simone de Beauvoir*, pp. 207–8.

36. See *Force of Circumstance*, p. 59; Bair, *Simone de Beauvoir*, p. 268; and Crosland, *Simone de Beauvoir*, p. 332.

37. See Sartre, "Forgers of myths," *Sartre on Theater*, pp. 33–43.

38. Hayman, *Writing Against*, p. 231.

39. *Force of Circumstance*, p. 63.

40. Ibid., pp. 63, 75.

41. For examples of the application of this theory, see Sartre's *Saint Genet* and *Critique of Dialectical Reason*, and Frantz Fanon, *The Wretched of the Earth*.

42. *The Ethics of Ambiguity*, p. 38.

43. *Force of Circumstance*, p. 77.

44. See ibid., p. 69, and Bair, *Simone de Beauvoir*, p. 316.

45. Cohen-Solal, *Sartre*, pp. 285, 286.

46. *Adieux*, p. 200.

47. Hayman, *Writing Against*, p. 236.

48. *Adieux*, p. 220.

49. Cohen-Solal, *Sartre*, p. 282.

50. *Force of Circumstance*, p. 86.

51. Hayman, *Writing Against*, p. 239. See Contat and Rybalka, *The Writings of Jean-Paul Sartre*, vol. 1, *A Bibliographical Life*, for information on Sartre's prodigious output during this period.

52. *Force of Circumstance*, p. 114.

53. Bair, *Simone de Beauvoir*, p. 316.

54. *Force of Circumstance*, pp. 113, 114–15, 116, 120.

55. Ibid., p. 131.

56. Ibid., pp. 132, 133.

57. *Letters to Sartre*, p. 414. Letter dated January 30 [1947].

58. Ibid., p. 423. Letter dated February 7, 1947.

59. Ibid., p. 415. Letter dated January 30 [1947].

60. Bair, *Simone de Beauvoir*, p. 329.

61. For information on Nelson Algren see Bettina Drew, *Nelson Algren: A life on the wild side*, Bloomsbury: London, 19?0, and H. E. F. Donohue, *Conversations with Nelson Algren*, Hill & Wang: New York, 1964.
62. *Letters to Sartre*, p. 434. Letter dated Friday, February 28 [1947].
63. Drew, *Nelson Algren*, pp. 376–7; Crosland, *Simone de Beauvoir*, p. 352.

Epilogue: Situations in Maturity

1. *Adieux*, p. 160.
2. For background, see Cohen-Solal, *Sartre*, pp. 294–311; Hayman, *Writing Against*, pp. 241–62; and Wilkinson, *The Intellectual Resistance in Europe*, pp. 95–106.
3. *Sartre by Himself*, p. 65
4. Cohen-Solal, *Sartre*, p. 309.
5. *Force of Circumstance*, p. 159. In *Adieux*, Beauvoir notes that she wrote her account of the RDR in *Force of Circumstance* partly under Sartre's dictation (see *Adieux*, p. 399).
6. *Force of Circumstance*, p. 157.
7. Ibid., pp. 186, 188.
8. *Adieux*, p. 397.
9. Jean-Paul Sartre, "Sartre talks of Beauvoir: An interview with Madeleine Gobeil," trans. Bernard Frechtman, *Vogue* (American edition), July 1965, p. 73.
10. In an interview with Hélène V. Wenzel in 1984, Beauvoir confirmed this association: "I really like my autobiobraphy because I feel that the ensemble of my autobiographical works wholly complements and completes my position on women." See "Interview with Simone de Beauvoir," *Yale French Studies*, no. 72, 1986, pp. 12–13. For Beauvoir's reflections on *The Second Sex* in the 1970s, see John Gerassi, "Simone de Beauvoir: *The Second Sex*: 25 years later," *Society*, January/February 1976, pp. 79–85.
11. For background, see *Force of Circumstance*, pp. 195–203.
12. Ibid., p. 197.
13. Bair, *Simone de Beauvoir*, p. 391.
14. On Colette Audry, see Dayan and Ribowska, *Simone de Beauvoir*, pp. 42–3, and Bair, *Simone de Beauvoir*, pp. 379–80. On Sartre encouraging Beauvoir to write about herself, see *Force of Circumstance*, p. 103, and Bair, *Simone de Beauvoir*, p. 325. On Beauvoir's feelings regarding female exceptionalism, see *Prime of Life*, pp. 62, 367, 572. On Sartre's later views of *The Second Sex*, see Sartre, *Life/Situations*, pp. 96–8ff.
15. Bair, *Simone de Beauvoir*, p. 364.
16. Ibid., p. 655.
17. Ibid., pp. 456–7; *Force of Circumstance*, p. 200.
18. In a letter to Algren, Beauvoir commented on her companion's predicament: "That nutty Sartre might just get shot if he refuses to marry her" (Francis and Gontier, *Simone de Beauvoir*, p. 294).
19. *Adieux*, pp. 227–8, 231.

20. *Force of Circumstance*, pp. 352–3.
21. Ibid., p. 381.
22. Ibid., p. 397.
23. *Adieux*, p. 367.
24. Hayman, *Writing Against*, pp. 350–1; *Force of Circumstance*, p. 581.
25. Hayman, *Writing Against*, p. 352.
26. *Force of Circumstance*, p. 674.
27. Ibid., p. 397.
28. *Sartre by Himself*, p. 77.
29. *Adieux*, pp. 318, 319, and Bair, *Simone de Beauvoir*, pp. 460, 666. Corydrane was a fashionable compound of aspirin and amphetamines. Sartre was also taking a drug called Orthédrine and drinking heavily at the time. His abuse of his body frightened Beauvoir. See Hayman, *Writing Against*, p. 315, and *Force of Circumstance*, p. 398.
30. Many of Sartre's uncompleted projects have been published posthumously.
31. Cohen-Solal, *Sartre*, p. 356.
32. Jean-Paul Sartre, *Critique of Dialectical Reason*, vol. 1, *Theory of Practical Ensembles*, trans. Alan Sheridan-Smith, ed. Jonathan Rée, NLB: London, 1976, p. 822.
33. Hayman, *Writing Against*, pp. 309–12, 320–1.
34. Danto, *Sartre*, pp. 112, 136.
35. *Adieux*, p. 422.
36. Cohen-Solal, *Sartre*, p. 375.
37. Hayman, *Writing Against*, p. 343.
38. *Adieux*, p. 371.
39. Simone de Beauvoir, *All Said and Done*, trans. Patrick O'Brian, Penguin: Harmondsworth, 1977, pp. 454–64.
40. Hayman, *Writing Against*, p. 393.
41. Francis and Gontier, *Simone de Beauvoir*, p. 327.
42. *Adieux*, p. 371.
43. Ibid., p. 397.
44. Hayman, *Writing Against*, p. 396.
45. On Beauvoir's declaration of willingness to be considered a feminist, see *All Said and Done*, p. 491. For additional information on *The Manifesto of the 343*, see Bair, *Simone de Beauvoir*, p. 547.
46. See Bair, *Simone de Beauvoir*, p. 547; Francis and Gontier, *Simone de Beauvoir*, pp. 337–8; and Schwarzer, *Simone de Beauvoir Today*, p. 13.
47. For an international collection of views of women on the importance of Beauvoir and her work to them, see Penny Forster and Imogen Sutton (eds), *Daughters of de Beauvoir*, The Women's Press: London, 1989.
48. Bair, *Simone de Beauvoir*, pp. 586–8.
49. *Adieux*, p. 3. For information on the period following Sartre's death, see Bair, *Simone de Beauvoir*, pp. 592–3ff.
50. Bair, *Simone de Beauvoir*, pp. 595–7.
51. Ibid., pp. 616–17.
52. *Force of Circumstance*, p. 275.

BIBLIOGRAPHY

Albérès, René, *Jean-Paul Sartre: Philosopher without faith*, trans. Wade Baskin (New York: Philosophical Library, 1961).

Appignanesi, Lisa, *Simone de Beauvoir* (Harmondsworth: Penguin, 1988).

Ascher, Carol, *Simone de Beauvoir: A life of freedom* (Hemel Hempstead: Harvester Wheatsheaf, 1981).

Bair, Deirdre, *Simone de Beauvoir: A biography*, (London: Jonathan Cape, 1990).

Bair, Deirdre, "Simone, Sartre and sex," *The Observer*, May 20, 1990, pp. 47–8.

Barnes, Hazel E., *The Literature of Possibility: A study in humanistic existentialism* (London: Tavistock, 1959).

Barnes, Hazel E., *Sartre* (London: Quartet, 1974).

Barnes, Hazel E., *Sartre and Flaubert* (London: University of Chicago Press, 1981).

Baruch, Elaine Hoffman, "The female body and the male mind: Reconsidering Simone de Beauvoir," *Dissent*, Summer 1987, pp. 351–8.

Beauvoir, Simone de, *Adieux: A farewell to Sartre*, trans. Patrick O'Brian (Harmondsworth: Penguin, 1985).

Beauvoir, Simone de, *All Men are Mortal*, trans. Leonard M. Friedman (London: Norton, 1992).

Beauvoir, Simone de, *All Said and Done*, trans. Patrick O'Brian (Harmondsworth: Penguin, 1977).

Beauvoir, Simone de, *America Day by Day*, trans. Patrick Dudley (London: Duckworth, 1952).

Beauvoir, Simone de, *Les Belles Images*, trans. Patrick O'Brian (London: Flamingo, 1985).

Beauvoir, Simone de, *The Blood of Others*, trans. Yvonne Moyse and Roger Senhouse (Harmondsworth: Penguin, 1964).

Beauvoir, Simone de, "Brigitte Bardot and the Lolita syndrome," *Esquire*, August 1959, pp. 32–8.

Beauvoir, Simone de, *The Ethics of Ambiguity*, trans. Bernard Frechtman, 1948 (rpt New York: Citadel Press, 1970).

Beauvoir, Simone de, *Force of Circumstance*, trans. Richard Howard (Harmondsworth: Penguin, 1968).

Beauvoir, Simone de, "Introduction" to Gisèle Halimi, *Djamila Boupacha*, trans. Peter Green (London: Andre Deutsch and Weidenfeld and Nicolson, 1962).

Beauvoir, Simone de, *Journal de guerre: septembre 1939–janvier 1941*, ed. Sylvie Le Bon de Beauvoir (Paris: Gallimard, 1990).

Beauvoir, Simone de, *Letters to Sartre*, trans. and ed. Quintin Hoare (London: Radius, 1991).

Beauvoir, Simone de, *Lettres à Sartre*, vol. I, *1930–1939*, vol. II, *1940–1963*, ed. Sylvie Le Bon de Beauvoir (Paris: Gallimard, 1990).

Beauvoir, Simone de, *The Long March*, trans. Austryn Wainhouse (London: Andre Deutsch and Weidenfeld and Nicolson, 1958).

Beauvoir, Simone de, *The Mandarins*, trans. Leonard M. Friedman (London: Flamingo, 1984).

Beauvoir, Simone de, *Memoirs of a Dutiful Daughter*, trans. James Kirkup (Harmondsworth: Penguin, 1963).

Beauvoir, Simone de, *Must We Burn de Sade?*, trans. Annette Michelson (London: Peter Nevill, 1953).

Beauvoir, Simone de, *Old Age*, trans. Patrick O'Brian (Harmondsworth: Penguin, 1977).

Beauvoir, Simone de, *The Prime of Life*, trans. Peter Green (Harmondsworth: Penguin, 1965).

Beauvoir, Simone de, "Pyrrhus and Cynéas," trans. Christopher Freemantle, *Partisan Review*, vol. 3, part 3, 1946, pp. 430–7.

Beauvoir, Simone de, *The Second Sex*, trans. H. M. Parshley, 1953 (rpt Harmondsworth: Penguin, 1972).

Beauvoir, Simone de, *She Came to Stay*, trans. Yvonne Moyse and Roger Senhouse (London: Flamingo, 1984).

Beauvoir, Simone de, *A Very Easy Death*, trans. Patrick O'Brian (Harmondsworth: Penguin, 1969).

Beauvoir, Simone de, "What love is – and isn't," *McCall's*, August 1965, pp. 71, 133.

Beauvoir, Simone de, *When Things of the Spirit Come First: Five early tales*, trans. Patrick O'Brian (London: Flamingo, 1983).

Beauvoir, Simone de, *The Woman Destroyed*, trans. Patrick O'Brian (London: Flamingo, 1984).

Brée, Germaine, *Camus and Sartre: Crisis and commitment* (London: Calder & Boyars, 1974).

Catalano, Joseph S., *A Commentary on Jean-Paul Sartre's Critique of Dialectical Reason*, vol. 1, *Theory of Practical Ensembles* (London: University of Chicago Press, 1986).

Catalano, Joseph S., *A Commentary on Sartre's Being and Nothingness* (London: University of Chicago Press, 1974).

Caws, Peter, *Sartre* (London: Routledge & Kegan Paul, 1979).

Celeux, Anne-Marie, *Jean-Paul Sartre, Simone de Beauvoir, une expérience commune, deux écritures* (Paris: Nizet, 1986).

Chiari, Joseph, *Twentieth-Century French Thought: From Bergson to Lévi-Strauss* (London: Paul Elek, 1975).

Cohen-Solal, Annie, *Sartre: A life* (London: Heinemann, 1987).

Collins, Douglas, *Sartre as Biographer* (London: Harvard University Press, 1980).

Collins, James, *The Existentialists: A critical study* (Chicago: Gateway, 1952).

Contat, Michel and Rybalka, Michel (eds), *The Writings of Jean-Paul Sartre*, vol. 1, *A Bibliographical Life*, trans. Richard C. McCleary (Evanston: Northwestern University Press, 1974).

Cottrell, Robert D., *Simone de Beauvoir* (New York: Frederick Ungar, 1975).

Craig, Carol, *Simone de Beauvoir's The Second Sex in the Light of the Hegelian Master–Slave Dialectic and Sartrian Existentialism*, PhD dissertation, University of Edinburgh, 1979.

Crosland, Margaret, *Simone de Beauvoir: The woman and her work* (London: Heinemann, 1992).

Danto, Arthur C., *Sartre*, 2nd edn, (London: Fontana, 1991).

David, Catherine, "Beauvoir elle-même," *Le Nouvel Observateur*, janvier 22, 1979, pp. 82–5, 88–9.

Dayan, Josée and Malka Ribowska, *Simone de Beauvoir* (Paris: Gallimard, 1979).

Donohue, H. E. F., *Conversations with Nelson Algren* (New York: Hill & Wang, 1964).

Drew, Bettina, *Nelson Algren: A life on the wild side* (London: Bloomsbury, 1990).

Duchen, Claire, *Feminism in France: From May '68 to Mitterand* (London: Routledge and Kegan Paul, 1986).

Evans, Mary, *Simone de Beauvoir: A feminist mandarin* (London: Tavistock, 1985).

Fallaize, Elizabeth, *The Novels of Simone de Beauvoir* (London: Routledge, 1988).

Fell, Joseph P., *Heidegger and Sartre: An essay on being and place* (New York: Columbia University Press, 1979).

Forster, Penny and Imogen Sutton (eds), *Daughters of de Beauvoir* (London: The Women's Press, 1989).

Francis, Claude and Fernande Gontier, *Simone de Beauvoir*, trans. Lisa Nesselson (London: Mandarin, 1989).

Francis, Claude and Janice Niepce, *Simone de Beauvoir et le cours du monde* (Paris: Klincksieck, 1978).

Gerassi, John, *Jean-Paul Sartre: Hated conscience of his century*, vol. 1, *Protestant or Protestor* (London: University of Chicago Press, 1989).

Gerassi, John, "Simone de Beauvoir: *The Second Sex*, 25 years later," *Society*, January/February 1976, pp. 79–85.

Goldthorpe, Rhiannon, *Sartre: Literature and theory* (Cambridge: Cambridge University Press, 1984).

Greene, Norman N., *Jean-Paul Sartre: The existentialist ethic* (Ann Arbor: University of Michigan Press, 1963).

Grene, Marjorie, *Sartre* (New York: New Viewpoints, 1973).

Grosz, Elizabeth, *Sexual Subversions: Three French feminists* (London: Allen & Unwin, 1989).

Hassan, Ihab, *The Dismemberment of Orpheus: Toward a postmodern literature*, 2nd edn (Madison: University of Wisconsin Press, 1982).

Hayman, Ronald, *Writing Against: A biography of Sartre* (London: Weidenfeld and Nicolson, 1986).

Heath, Jane, *Simone de Beauvoir* (Hemel Hempstead: Harvester Wheatsheaf, 1989).

Heidegger, Martin, *Being and Time*, trans. John Macquarrie and Edward Robinson (Oxford: Blackwell, 1962).

Hemingway, Ernest, *A Farewell to Arms* (New York: Scribner's, 1929).

Houston, Mona Tobin, "The Sartre of Madame de Beauvoir," *Yale French Studies*, no. 30, 1964, pp. 23–9.

Howells, Christina (ed.), *The Cambridge Companion to Sartre* (Cambridge: Cambridge University Press, 1992).

Jameson, Fredric (ed.), "Sartre after Sartre," *Yale French Studies*, no. 68, 1985.

Jameson, Fredric, *Sartre: The origins of a style* (London: Yale University Press, 1961).

Jardine, Alice, "Interview with Simone de Beauvoir," *Signs*, vol. 5, no. 2, Winter 1979, pp. 224–36.

Jouve, Nicole Ward, "How *The Second Sex* stopped my aunt from watering the horse-chestnuts: Simone de Beauvoir and contemporary feminism," *White Woman Speaks with Forked Tongue: Criticism as autobiography* (London: Routledge, 1991).

Keefe, Terry, *Simone de Beauvoir: A study of her writings* (Totowa, NJ: Barnes & Noble, 1983).

Kern, Edith (ed.), *Sartre* (Englewood Cliffs, NJ: Prentice Hall, 1962).

LaCapra, Dominick, *A Preface to Sartre* (London: Methuen, 1979).

Laing, R. D. and D. G. Cooper, *Reason and Violence: A decade of Sartre's philosophy, 1950–1960* (London: Tavistock, 1964).

Lawson, Sylvia, "All in the family," *London Review of Books*, December 3, 1992, pp. 15–16.

Le Doeuff, Michèle, *Hipparchia's Choice: An essay concerning women, philosophy, etc.*, trans. Trista Selous (Oxford: Blackwell, 1991).

Le Doeuff, Michèle, "Operative philosophy: Simone de Beauvoir and existentialism," trans. Colin Gordon, *Ideology and Consciousness*, no. 6, Autumn 1979, pp. 47–57.

Le Doeuff, Michèle, "Women and philosophy," trans. Debby Pope, *Radical Philosophy*, no. 17, 1977, pp. 2–11.

Lloyd, Genevieve, "Masters, slaves and others," *Radical Philosophy*, no. 34, Summer 1983, pp. 2–9.

Lloyd, Genevieve, *The Man of Reason: "Male" and "female" in western philosophy* (London: Methuen, 1984).

McCall, Dorothy, "Simone de Beauvoir, *The Second Sex*, and Jean-Paul Sartre," *Signs*, vol. 5, no. 2, Winter 1979, pp. 209–23.

McCall, Dorothy, *The Theatre of Jean-Paul Sartre* (London: Columbia University Press, 1969).

Macquarrie, John, *Existentialism* (Harmondsworth: Penguin, 1973).

Madsen, Axel, *Hearts and Minds: The common journey of Simone de Beauvoir and Jean-Paul Sartre* (New York: William Morrow, 1977).

Manser, Anthony, *Sartre: A philosophic study* (New York: Oxford University Press, 1966).

Marks, Elaine (ed.), *Critical Essays on Simone de Beauvoir* (Boston: G. K. Hall, 1987).

Marks, Elaine, *Simone de Beauvoir: Encounters with death* (New Brunswick, NJ: Rutgers University Press, 1973).

Marks, Elaine and Isabelle de Courtivron (eds), *New French Feminisms* (Hemel Hempstead: Harvester Wheatsheaf, 1981).

Merleau-Ponty, Maurice, *Phenomenology of Perception*, trans. Colin Smith (London: Routledge, 1962).

Mészáros, István, *The Work of Sartre*, vol. 1, *Search for Freedom* (Atlantic Highlands, NJ: Humanities Press, 1979).

Midgley, Mary and Judith Hughes, *Women's Choices: Philosophical problems facing feminism* (London: Weidenfeld and Nicolson, 1983).

Moi, Toril, *Feminist Theory and Simone de Beauvoir* (London: Blackwell, 1990).

Moi, Toril, *Sexual/Textual Politics: Feminist literary theory* (London: Methuen, 1985).

Murdoch, Iris, *Sartre: Romantic rationalist*, 1953 (rpt London: Fontana, 1967).

Okely, Judith, *Simone de Beauvoir: A re-reading*, (London: Virago, 1986).

Radford, C. B., "Simone de Beauvoir: Feminism's friend or foe?," *Nottingham French Studies*, vol. 6, no. 2, October 1967, pp. 87–102; vol. 7, no. 1, May 1968, pp. 39–53.

Roubiczek, Paul, *Existentialism: For and against* (Cambridge: Cambridge University Press, 1964).

Russell, Bertrand, *The Problems of Philosophy*, 1912 (rpt London: Oxford University Press, 1967).

Sartre, Jean-Paul, *The Age of Reason*, trans. Eric Sutton (New York: Bantam, 1959).

Sartre, Jean-Paul, *Altona, Men without Shadows, The Flies*, trans. Sylvia and George Leeson, Kitty Black, and Stuart Gilbert (Harmondsworth: Penguin, 1962).

Sartre, Jean-Paul, *Anti-Semite and Jew*, trans. George J. Becker (New York: Schocken Books, 1965).

Sartre, Jean-Paul, *Baudelaire*, trans. Martin Turnell, 1949 (rpt London: Hamish Hamilton, 1964).

Sartre, Jean-Paul, *Being and Nothingness: An essay on phenomenological ontology*, trans. Hazel E. Barnes (New York: Philosophical Library, 1956).

Sartre, Jean-Paul, *The Chips are Down*, trans. Louise Verèse (London: Rider, 1951).

Sartre, Jean-Paul, *Crime passionnel*, trans. Kitty Black (London: Methuen, 1961).

Sartre, Jean-Paul, *Critique of Dialectical Reason*, vol. 1, *Theory of Practical Ensembles*, trans. Alan Sheridan-Smith, ed. Jonathan Rée (London: NLB, 1976).

Sartre, Jean-Paul, *The Emotions: Outline of a theory*, trans. Bernard Frechtman (New York: Philosophical Library, 1948).

Sartre, Jean-Paul, *Essays in Existentialism*, ed. Wade Baskin (New York: Citadel Press, 1990).

Sartre, Jean-Paul, *Existentialism*, trans. Bernard Frechtman (New York: Philosophical Library, 1947).

Sartre, Jean-Paul, *Existentialism and Humanism*, trans. Philip Mairet (London: Methuen, 1948).

Sartre, Jean-Paul, *The Freud Scenario*, trans. Quintin Hoare, ed. J.-B. Pontalis (London: Verso, 1985).

Sartre, Jean-Paul, *In Camera and Other Plays*, trans. Kitty Black and Stuart Gilbert (Harmondsworth: Penguin, 1982).

Sartre, Jean-Paul, *In the Mesh: A scenario*, trans. Mervyn Small (London: Andrew Dakers, 1954).

Sartre, Jean-Paul, *Intimacy and Other Stories*, trans. Lloyd Alexander (New York: New Directions, 1948).

Sartre, Jean-Paul, *Iron in the Soul*, trans. Gerard Hopkins (Harmondsworth: Penguin, 1963).

Sartre, Jean-Paul, "Jean-Paul Sartre: A candid conversation with the charismatic fountainhead of existentialism and rejector of the Nobel Prize," *Playboy*, May 1965, pp. 69–76.

Sartre, Jean-Paul, *Lettres au Castor et à quelques autres*, vol. 1, *1926–1939*, vol. 2, *1940–1963*, ed. Simone de Beauvoir (Paris: Gallimard, 1983).

Sartre, Jean-Paul, *Life/Situations: Essays written and spoken*, trans. Paul Auster and Lydia Davis (New York: Pantheon, 1977).

Sartre, Jean-Paul, *Literary and Philosophical Essays*, trans. Annette Michelson (New York: Collier, 1962).

Sartre, Jean-Paul, *Mallarmé: Or the poet of nothingness*, trans. Ernest Sturm (London: Pennsylvania State University Press, 1988).

Sartre, Jean-Paul, *Nausea*, trans. Robert Baldick (Harmondsworth: Penguin, 1965).

Sartre, Jean-Paul, *Politics and Literature*, trans. J. A. Underwood and John Calder (London: Calder & Boyars, 1973).

Sartre, Jean-Paul, "Preface" to Frantz Fanon, *The Wretched of the Earth*, trans. Constance Farrington (Harmondsworth: Penguin, 1967).

Sartre, Jean-Paul, *The Psychology of Imagination*, trans. Bernard Frechtman (New York: Washington Square Press, 1966).

Sartre, Jean-Paul, *The Reprieve*, trans. Eric Sutton (Harmondsworth: Penguin, 1963).

Sartre, Jean-Paul, *Saint Genet: Actor and martyr*, trans. Bernard Frechtman (New York: New American Library, 1963).

Sartre, Jean-Paul, "Sartre et les femmes: un entretien avec Catherine Chaine," *Le Nouvel Observateur*, no. 638, janvier 30–février 6, 1977.

Sartre, Jean-Paul, *Sartre by Himself*, film directed by Alexandre Astruc and Michel Contat, trans. Richard Seaver (New York: Urizen, 1978).

Sartre, Jean-Paul, *Sartre on Theater*, ed. Michel Contat and Michel Rybalka, trans. Frank Jellinek (London: Quartet, 1976).

Sartre, Jean-Paul, "Sartre talks of Beauvoir: An interview with Madeleine Gobeil," trans. Bernard Frechtman, *Vogue* (American edition), July 1965, pp. 72–3.

Sartre, Jean-Paul, *Search for a Method*, trans. Hazel E. Barnes (New York: Vintage, 1968).

Sartre, Jean-Paul, *Three Plays: Kean, Nekrassov, The Trojan Women*, trans. Kitty Black, Sylvia and George Leeson, and Ronald Duncan (Harmondsworth: Penguin, 1969).

Sartre, Jean-Paul, *The Transcendence of the Ego: An existentialist theory of consciousness*, trans. Forrest Williams and Robert Kirkpatrick (New York: Noonday Press, 1957).

Sartre, Jean-Paul, *War Diaries: Notebooks from a phoney war, November 1939–March 1940*, trans. Quintin Hoare (London: Verso, 1984).

Sartre, Jean-Paul, *What is Literature?*, trans. Bernard Frechtman (London: Methuen, 1950).

Sartre, Jean-Paul, *Witness to my Life: The letters of Jean-Paul Sartre to Simone de Beauvoir, 1926–1939*, ed. Simone de Beauvoir, trans. Lee Fahnestock and Norman MacAfee (New York: Charles Scribner's, 1992).

Sartre, Jean-Paul, *The Words*, trans. Bernard Frechtman (New York: George Braziller, 1964).

Sartre, Jean-Paul, *The Writings of Jean-Paul Sartre*, vol. 2, *Selected Prose*, ed. Michel Contat and Michel Rybalka, trans. Richard McCleary (Evanston, IL: Northwestern University Press, 1974).

Schroeder, William Ralph, *Sartre and His Predecessors: The self and the other* (London: Routledge and Kegan Paul, 1984).

Schwarzer, Alice, *Simone de Beauvoir Today: Conversations 1972–1982*, trans. Marianne Howarth (London: Chatto & Windus, 1984).

Scriven, Michael, *Sartre's Existential Biographies* (London: Macmillan, 1984).

Showalter, Elaine, *Sexual Anarchy: Gender and culture at the fin de siècle* (London: Bloomsbury, 1991).

Siegel, Liliane, *In the Shadow of Sartre*, trans. Barbara Wright (London: Collins, 1990).

Silverman, Hugh J. and Frederick A. Elliston, *Jean-Paul Sartre: Contemporary approaches to his philosophy* (Hemel Hempstead: Harvester Wheatsheaf, 1980).

Simons, Margaret A., "Beauvoir and Sartre: The philosophical relationship," *Yale French Studies*, no. 72, 1986, pp. 165–79.

Simons, Margaret A., "The silencing of Simone de Beauvoir: Guess what's missing from *The Second Sex*," *Women's Studies International Forum*, vol. 6, no. 5, 1983, pp. 559–64.

Singer, Linda, "Interpretation and retrieval: Rereading Beauvoir," *Women's Studies International Forum*, vol. 8, no. 3, 1985, pp. 231–8.

Sontag, Susan, "Sartre's *Saint Genet*" in *Against Interpretation and Other Essays* (New York: Delta, 1961).

Thody, Philip, *Jean-Paul Sartre* (London: Macmillan, 1992).

Thody, Philip, *Sartre: A biographical introduction* (London: Studio Vista, 1971).

Walters, Margaret, "The rights and wrongs of women: Mary Wollstonecraft, Harriet Martineau, Simone de Beauvoir" in *The Rights and Wrongs of Women*, ed. Juliet Mitchell and Ann Oakley (Harmondsworth: Penguin, 1976).

Warnock, Mary, *Existentialism* (Oxford: Oxford University Press, 1970).

Warnock, Mary, *The Philosophy of Sartre* (London: Hutchinson, 1965).

Wenzel, Hélène Vivienne (ed.), "Simone de Beauvoir: Witness to a century", *Yale French Studies*, no. 72, 1986.

Whitmarsh, Anne, *Simone de Beauvoir and the Limits of Commitment* (Cambridge: Cambridge University Press, 1981).

Wilcox, Helen, Keith McWatters, Ann Thompson, and Linda R. Williams (eds), *The Body and the Text: Hélène Cixous, reading and teaching* (Hemel Hempstead: Harvester Wheatsheaf, 1990).

Wilkinson, James D., *The Intellectual Resistance in Europe* (London: Harvard University Press, 1981).

Winegarten, Renée, *Simone de Beauvoir: A critical view* (Oxford: Berg, 1988).

Woodward, Kathleen, "Simone de Beauvoir: Aging and its discontents" in *The Private Self: Theory and practice of women's autobiographical writings*, ed. Shari Benstock (London: Routledge, 1988), pp. 90–114.

Young-Bruehl, Elisabeth, *Mind and the Body Politic* (London: Routledge, 1989).

INDEX